Contents

REALITY CHECK

The Nature and Performance of Voluntary Environmental Programs in the United States, Europe, and Japan

EDITED BY
RICHARD D. MORGENSTERN
AND WILLIAM A. PIZER

RESOURCES FOR THE FUTURE
WASHINGTON, DC, USA

Printed in the United States of America

No part of this publication may be reproduced by any means, whether electronic or mechanical, without written permission. Requests to photocopy items for classroom or other educational use should be sent to the Copyright Clearance Center, Inc., Suite 910, 222 Rosewood Drive, Danvers, MA 01923, USA (fax +1 978 646 8600; www.copyright.com). All other permissions requests should be sent directly to the publisher at the address below.

An RFF Press book
Published by Resources for the Future
1616 P Street, NW
Washington, DC 20036–1400
USA
www.rffpress.org

Library of Congress Cataloging-in-Publication Data

Morgenstern, Richard D.
  Reality check : the nature and performance of voluntary environmental programs in the United States, Europe, and Japan / Richard D. Morgenstern, William A. Pizer.
    p. cm.
  Includes bibliographical references and index.
  ISBN 1-933115-36-X (cloth : alk. paper) – ISBN 1-933115-37-8 (pbk. : alk. paper)
  1. Environmental policy–United States–Citizen participation–Case studies.
  2. Environmental policy–Europe–Citizen participation–Case studies. 3. Environmental policy–Japan–Citizen participation–Case studies. 4. Voluntarism–United States–Case studies. 5. Voluntarism–Europe–Case studies. 6. Voluntarism–Japan–Case studies. 7. Social responsibility of business–United States–Case studies.
  8. Social responsibility of business–Europe–Case studies. 9. Social responsibility of business–Japan–Case studies.
  I. Morgenstern, Richard D. II. Title.
  GE180.P585 2007
  363.7'0525–dc22   2006036837

The paper in this book meets the guidelines for permanence and durability of the Committee on Production Guidelines for Book Longevity of the Council on Library Resources. This book was typeset by Peter Lindeman. It was copyedited by Patti Miller. The cover was designed by Marc Meadows. Cover image by Getty Images.

ISBN 1-933115-36-X (cloth)   ISBN 1-933115-37-8 (paper)

About Resources for the Future *and* RFF Press

RESOURCES FOR THE FUTURE (RFF) improves environmental and natural resource policymaking worldwide through independent social science research of the highest caliber. Founded in 1952, RFF pioneered the application of economics as a tool for developing more effective policy about the use and conservation of natural resources. Its scholars continue to employ social science methods to analyze critical issues concerning pollution control, energy policy, land and water use, hazardous waste, climate change, biodiversity, and the environmental challenges of developing countries.

RFF PRESS supports the mission of RFF by publishing book-length works that present a broad range of approaches to the study of natural resources and the environment. Its authors and editors include RFF staff, researchers from the larger academic and policy communities, and journalists. Audiences for publications by RFF Press include all of the participants in the policymaking process—scholars, the media, advocacy groups, NGOs, professionals in business and government, and the public.

Preface

The idea for this volume originated in a conversation we had in 2003 about the lack of objective evaluations of voluntary programs that encourage firms to improve their environmental performance. Governments around the world are increasingly turning to these policies, especially for the control of greenhouse gas emissions. Yet the majority of work either focuses on participation (rather than outcomes) or tends to be written by interested stakeholders. As we began to organize our thinking for a book on the evaluation of voluntary environmental programs, we were skeptical that this evaluation could center on quantitative analysis given our expectation that data on environmental outcomes would be hard to measure. Therefore, we chose early on to follow a case study approach with the expectation that the evaluation would be mostly qualitative. We identified highly respected scholars who have puzzled over these same issues, had prior familiarity with individual programs and, it turns out, were excited to join in our effort. What has emerged is a detailed and surprisingly quantitative set of assessments of an issue that has heretofore been dominated by stakeholder rhetoric rather than evidence.

As a disclaimer we note that we both have personal associations with voluntary programs dating back to our experiences in government. Morgenstern first worked on such programs while serving in the U.S. EPA's Policy Office in the early 1990's and was closely involved in the development of the Climate Wise program that is the subject of Chapter 7. During his 2001–2002 service on the Council of Economic Advisers, Pizer participated in the consideration of various voluntary approaches in climate change policy and worked on strengthening the Department of Energy's voluntary emission registry.

Many different people contributed to this volume. First and foremost, we thank the case study authors. In addition, we are grateful to the individuals who served as peer reviewers for the different chapters at an authors' workshop that was held at RFF in December 2005: Joe Aldy, Terry Dinan, Kathleen Hogan, Andreas Lange, Tom Lyon, Janet Peace, and Bob Shackleton. Three anonymous referees also provided valuable comments on the entire manuscript, as well as our indefatigable publisher at RFF Press, Don Reisman, and his production team.

This project would not have happened without the generous support of the Smith Richardson Foundation and the guidance of our program officer, Mark Steinmeyer. As is always the case with our joint work, Devra Davis offered many informal and invaluable insights while providing Morgenstern with moral support that we both appreciate.

RICHARD D. MORGENSTERN
WILLIAM A. PIZER

Contributors

Christoph Böhringer supervised the energy economics research group at the Institute of Energy Economics, University of Stuttgart, Germany, from 1995 to 1998. From 1999 to 2006 he was head of the Environmental and Resource Economics, Environmental Management Department at the Centre for European Economic Research, Mannheim. He holds the chair in economic policy at the University of Oldenburg. His research focuses on the impact analysis of environmental, energy, and trade policies.

Gildas de Muizon is CEO of Microeconomix, a consulting company in economics applied to legal disputes and environmental issues.

Manuel Frondel is research coordinator and head of the Environment and Resources Department at RWI, Rheinisch-Westfaelisches Institut fuer Wirtschaftsforschung, Essen, Germany. He was previously a research fellow at the Centre for European Economic Research (ZEW), Mannheim, and part-time professor at the University of Applied Sciences, Heilbronn, Germany. His research interests include applied econometrics in environmental, resource, and energy economics. Frondel has published in international journals such as *The Review of Economics and Statistics, Economics Letters*, and *The Energy Journal.*

Matthieu Glachant is a research fellow in environmental economics at the Ecole des mines de Paris (CERNA). His research focuses on the political economy of voluntary instruments and the economics of waste. He is the coauthor of *Voluntary Approaches for Environmental Policy: An Assessment.*

Madhu Khanna is a professor in the Department of Agricultural and Consumer Economics at the University of Illinois at Urbana-Champaign. Her recent research analyzes corporate environmental behavior and the implications of environmental management systems and information disclosure policies for environmental protection. She is a University of Illinois scholar and a teaching fellow of the North American Colleges and Teachers of Agriculture.

Signe Krarup is a senior research fellow at akf, Institute of Local Government Studies, Denmark. Her main fields of research are in environmental economics, monitoring and enforcement strategies and the demand for organic foods. Recent publications include the book *Environment, Information and Consumer Behaviour: An Introduction* (with C. S. Russell as coeditor), and an article in *Journal of Regulatory Economics* (with C. S. Russell and L.G. Hansen as co-authors).

Katrin Millock is a senior research fellow at Centre d'Economie de la Sorbonne (CNRS-University of Paris) specializing in environmental and resource economics and agricultural policy. Her recent research projects focus on incentives for technology adoption, ex post evaluation of environmental taxes, economic instruments and social norms, and consumer information and food safety.

Richard D. Morgenstern is a senior fellow at Resources for the Future (RFF) whose research focuses on the design of environmental policies, including economic incentive measures, in relation to regulatory issues and climate change policies. Previously, he directed the U.S. Environmental Protection Agency's Office of Policy Analysis, and he has acted as the EPA's assistant administrator for policy, planning and evaluation, and as the agency's deputy administrator. He also served as a senior economic counselor to the under secretary of state for global affairs at the U.S. State Department. He is editor or coeditor of three other RFF books: *Economic Analyses at EPA: Assessing Regulatory Impact* (1997); *Choosing Environmental Policy* (2004); and *New Approaches on Energy and the Environment* (2004).

William A. Pizer is a senior fellow at Resources for the Future (RFF), where his research seeks to quantify how the design of environmental policy affects costs and effectiveness, including questions about uncertainty, technological change, flexibility mechanisms, and valuation over long-time horizons. Since August 2002, Pizer has worked part-time as a senior economist at the National Commission on Energy Policy; during 2001–2002, he served as a senior economist at the President's Council of Economic Advisers. He serves on a variety of advisory panels and is a lead author on the *Intergovernmental Panel on Climate Change Fourth Assessment Report.*

Alan H. Sanstad is a staff scientist at the Lawrence Berkeley National Laboratory and a research affiliate of the Energy and Resources Group, University of California at Berkeley. His recent research has focused upon the economics and policy analysis of energy efficiency and greenhouse gas mitigation. He serves as an adviser to the California Energy Commission.

Jhih-Shang Shih is a fellow at Resources for the Future who specializes in environmental management and policy analysis. His research projects focus on voluntary programs, renewable energy, cost and benefit of air pollution control, recycling, and community water systems. His articles have appeared in the *Review of Economics and Statistics*, the *Journal of Environmental Economics and Management*, and the *Journal of Environmental Management*, among others.

Taishi Sugiyama is the leader of the Climate Policy Project at the Central Research Institute of Electric Power Industry (CRIEPI), Japan. He serves as a lead author of the *Intergovernmental Panel on Climate Change Fourth Assessment Report*, a member of the Future Framework Committee that made recommendations to the Japanese government on a post-2012 framework, and as a member of the Market Mechanisms Committee of the Japanese government. He served on the Small Scale Clean Development Mechanism (CDM) Panel of the CDM Executive Board and the Scientific Steering Committee of the International Human Dimension Program (IHDP)/ Institutional Dimensions of Global Environmental Change (IDGEC).

Masayo Wakabayashi is a researcher at the Central Research Institute of Electric Power Industry (CRIEPI), Japan. Her recent research focuses on case studies of environmental policy measures, as represented by regulatory policy and industrial responses.

REALITY CHECK

*The Nature and Performance
of Voluntary Environmental
Programs in the United States,
Europe, and Japan*

1

Introduction

The Challenge of
Evaluating Voluntary Programs

Richard D. Morgenstern and William A. Pizer

The explosive growth in voluntary environmental programs since the early 1990s in the United States, Europe, and Japan reflects, in part, changing societal attitudes about the environment and a growing optimism on the possibility of enhanced cooperation between government and business. It also reflects the widespread frustration with the long and expensive battles often associated with new environmental regulations. In most cases, voluntary programs are being used to control pollutants that have not yet been regulated and for which legislative authority may be difficult to obtain. Unlike market-based approaches to environmental management, where the conceptual roots are largely academic, voluntary programs have emerged as a pragmatic response to the need for more flexible ways to protect the environment.

The key question considered in this volume is whether these programs actually work as advertised. That is, do voluntary programs deliver the promise of significant environmental gains without the burdens associated with mandatory regulation? Do they improve environmental and conservation outcomes relative to a realistic baseline, or do they pave the way for other actions that do? Quantitatively, how large are the likely gains? Can such approaches serve as a substitute for mandatory requirements or should only modest gains be expected from these efforts?

The existing literature—which primarily emphasizes the motivation of firms to participate rather than the environmental accomplishments of the programs—provides only limited answers to the questions above. Most studies have focused on the 33/50 program in the United States, which is aimed at cutting toxic releases. A unique feature of this program, which facilitates performance evaluation, is that relevant emissions data from the Toxics Release Inventory (TRI) are routinely collected for both participating and non-participating firms. While this volume contains an updated analysis of 33/50 as well, the remaining six of seven case studies address carbon- or energy-related programs. This intentional focus reflects both the increasing attention given to voluntary climate change and energy-efficiency programs, as well as the limited existing work in this area.

Although there are many types of voluntary programs—often grouped by whether the parameters of the program originate in the private sector, the public sector, or are negotiated between the two—all of them involve some form of commitment to improve environmental performance beyond the existing legal requirements. Government often is directly involved in setting the goals and monitoring the results, although that need not be the case.

The growing popularity of these programs is undeniable. A 2005 survey identified 87 voluntary programs at the U.S. Environmental Protection Agency (EPA), up from 54 in 1999 and 28 in 1996 (U.S. EPA 2005). In fiscal year 2006, voluntary programs comprised 1.6% of EPA's operating budget. Dozens more programs operate at the U.S. Department of Energy and other federal agencies and at the state level. Issues covered include climate change, energy, waste, water, toxics, and agriculture. In Europe, voluntary programs also address a variety of environmental issues, although most are focused on climate change. The European Environment Agency (1997) reported that more than 300 voluntary agreements have been established between national governments and industry associations, with the Netherlands and Germany accounting for about two-thirds of them. In Japan, the focus is on so-called self-regulating mechanisms, mostly involving single firms working with local agencies (Baranzini and Thalmann 2004). A 1999 paper reported that Japan has more than 30,000 such arrangements (Tsutumi 1999). The largest program is the Action Plan on the Environment, initiated in 1997 by the Keidanren, an association of large firms, mostly in energy-intensive sectors.

In principle, voluntary programs offer opportunities for business to get hands-on experience with new types of environmental problems without the straightjacket of regulation and, in the process, to enhance their environmental reputation with government, customers, investors, communities, employees, and other firms. In some cases, the firms' participation may represent an effort to shape future regulations or to stave off mandatory requirements altogether. Some or all of these benefits may be reflected in the firms' bottom line over the short or long term.

Voluntary programs also provide opportunities for government agencies to gain experience with new problems and new industries. Most importantly, they provide opportunities to achieve environmental improvements more quickly, and with lower administrative costs, than otherwise would be possible and, sometimes, via more holistic approaches than the media-specific, end-of-pipe focus of most existing legislation. In the view of some observers, by encouraging proactive approaches from industry, voluntary programs may help foster a common understanding of both environmental problems and the mutual responsibilities to address them.[1]

Notwithstanding the many potential benefits of voluntary approaches, the absence of deliberate price or regulatory signals to encourage fundamental changes in corporate or consumer actions or to stimulate demand for cleaner technologies is a clear limitation. The term "regulatory capture" applies when the targets established for the voluntary programs reflect only a business-as-usual scenario. Free riding, wherein some firms avoid any effort while other, proactive firms voluntarily address a problem and keep further regulation at bay, may be an issue with certain voluntary programs. Taking this a step further, voluntary approaches may represent a shift in emphasis from the "worst" polluters to those most willing to abate on their own initiative. Some, particularly in the environmental community, see voluntary programs as a distraction from the real work of taking mandatory action.[2]

Since business is inherently dynamic, with firms constantly confronting new challenges, opportunities, and technologies, it is not sufficient simply to look at two distinct points in time to see if firms' environmental performance has improved. Rather, environmental gains must be assessed with reference to a credible representation of what would have happened otherwise. Defining such a baseline is, of course, quite difficult to do. One approach is to construct a business-as-usual forecast using the best available data. However, such an approach is limited by the large number of unpredictable influences on outcomes. An alternative is to compare participants to a suitably chosen group of nonparticipants. Still, biases may arise if participants and nonparticipants differ in some systematic way—for example, if participants are bigger, faster-growing, or better managed. Unless the comparisons are carefully constructed, observed differences between participants and nonparticipants may reflect factors other than the effects of the program.

If we imagined a laboratory setting, the most transparent way to measure the environmental performance of a voluntary program—or any program—would be to conduct a scientific experiment to see whether firms randomly assigned to the program exhibited different outcomes than those randomly assigned to a control group. Because the two groups would be otherwise identical (due to randomization), this would yield an unbiased estimate of the effect of the voluntary program on environmental performance. In real life, we rarely see such randomized experiments and are instead left with either forecast baselines or imperfect control groups. This provides only limited evidence on the environmental performance of participating firms compared to what realistically would have happened otherwise.

Getting credible answers to these questions is important. Protagonists and antagonists of the trend toward voluntary approaches are increasingly at odds, sometimes drawing opposite conclusions about the same program. Protagonists, typically on the side of industry, see voluntary programs as a more practical, flexible approach to regulation. Antagonists, including some environmental advocates, often see voluntary programs as an obstacle to more stringent, mandatory programs. This polarization may be partly a consequence of poor information. While intuition and anecdotes may provide some reason for believing that a given program has or has not had a beneficial environmental impact, careful empirical analysis with peer review is much more convincing. The goal of this volume is to help fill that void.

The Existing Literature

The literature on voluntary programs contains a variety of descriptors to identify particular mechanisms: self-regulation, negotiated agreements, environmental covenants, business-led environmental strategies, and others. Nonetheless, a loose taxonomy has evolved, with three reasonably distinct bins based on how the parameters of the commitment are determined:

- *Unilateral agreements by industrial firms.* Business-led corporate programs fall under this heading, as do commitments or reduction targets chosen by firms or industry associations. Examples of such agreements in the United States include the Chemical Manufacturers Association's "Responsible Care" program for reducing chemical hazards and McDonald's replacement of its Styrofoam clamshell containers with paper packaging. In Germany, the Industry and Trade Association's plan to reduce carbon dioxide emissions also was a unilateral agreement. The study in this volume of the German cement industry is part of that plan.
- *Public voluntary programs.* Participating firms agree to protocols that have been developed by environmental agencies or other public bodies. Although the public agencies may promote the programs to industry, they generally do not negotiate over the specific terms. Eligibility criteria, rewards, obligations, and other elements are established by the public agencies. Examples of such programs in the United States include the 33/50 program, which focused on toxics, and Climate Wise, which focused on greenhouse gases, and the residential demand-side management (DSM) programs operated in California.[3] All three of these programs are subject to in-depth examination in this volume.
- *Negotiated agreements.* Consisting of a target and timetable for attaining the agreed-upon environmental objectives, these are created out of a negotiation between government authorities and a firm or industry group over specific terms. In some cases, participating firms receive relief from an otherwise burdensome tax, making the voluntary notion of the program somewhat hazy. In many countries, firms are held liable for compliance on an individual basis while in others, such as Japan, industries generally are liable on a collective basis for the environmental performance stipulated in the agree-

ments. Examples of negotiated agreements, which are most frequently used in Europe and Japan, include the Danish program on industrial energy efficiency, the Voluntary Climate Agreements in the United Kingdom, and the Keidanren Voluntary Action Plan on the Environment in Japan, all three of which are examined in this volume. The French agreement on the treatment of the end-of-life of vehicles and the U.S. XL program are further examples of negotiated agreements.

It is worth noting that while the delineation of voluntary programs into these three categories may seem clear cut, virtually all of these programs involve some degree of dialogue between government and firms over various terms. The dialogue may be indirect or informal and may feature stronger or weaker positions on each side. Hence, these categories highlight whether the terms are overwhelmingly determined by either firms or government with only informal dialogue, as in the first two categories, or whether there is more give-and-take and typically more formal interactions, as in the latter.

Economic analysis suggests that since environmental mitigation typically is not costless and the benefits not appropriable by the firm, profit-maximizing firms have little incentive to undertake such activities unless mandated by government to do so. It is not surprising, therefore, that as measured by the number of articles or books published on the subject, by far the dominant issue in the academic literature on voluntary programs concerns the motivation for firms to participate in the programs. Extensive theoretical and some empirical work has focused on the importance of preempting regulatory threats; the potential to influence future regulations; the effects on stakeholder relations and the firms' public image; the importance (or unimportance) of technical assistance and financial incentives to the firms' participation decision; the economic efficiency of the programs; the role of competitive pressures; and the potential to bring about savings in transaction or compliance costs. Several studies have shown the importance of public recognition provided by participation in a voluntary program to be a key motivation for firms.[4]

While the literature on the motivation for firms to participate in voluntary programs is extensive, there are only a limited number of previous analyses of environmental performance. The largely theoretical work on the issue suggests that participation in voluntary programs does not guarantee an improvement in actual performance. While it may encourage the exchange of information about best practices, a key factor may be to provide insurance to firms against stakeholder pressure. Thus, by implication, it might be argued that participation in voluntary programs may actually reduce incentives to cut emissions if it is successful in staving off stakeholder pressure for more stringent actions. Theoretical studies have shown that improvements in actual environmental performance depend on the extent to which voluntary programs lead to lower abatement costs relative to mandatory regulation; the likelihood that regulation will be imposed even if the program is not effective; the extent to which the regulator is willing to subsidize pollution reduction; the willingness of consumers to pay for green products; and other factors.[5]

In considering environmental performance of voluntary programs it is useful to distinguish between those programs that focus on the adoption of particular technologies (e.g., Green Lights, now part of the Energy Star Program) and those that focus directly on environmental performance (e.g., 33/50, Climate Wise, or various audit-based programs). In the former, success is measured as adoption of specific technologies. In the latter, it is measured as a reduction in emissions. In both cases, there is the need to define a baseline: Measured over the same period, how many firms (or households) would have installed the technologies, or how much would emissions have been reduced, even without the voluntary program?[6]

Technology programs can be difficult to evaluate because of the general absence of comprehensive databases on the performance of facilities that have not adopted the particular technologies. Despite this limitation, a number of these programs have been subject to at least some evaluation. The Green Lights Program is an innovative, voluntary, pollution-prevention program sponsored by the U.S. EPA focused on the installation of energy-efficient lighting where profitable and where lighting quality can be maintained or improved. DeCanio (1998) finds that the energy-efficiency investments carried out under this program yielded annual real rates of return averaging 45%. DeCanio and Watkins (1998) find that specific characteristics of firms affect their decision to join Green Lights and commit to a program of investments in lighting efficiency.[7]

Energy Star is a voluntary labeling program designed to identify and promote energy-efficient products to reduce greenhouse gas emissions. Dowd et al. (2001) cite specific product-purchase decisions being influenced by Energy Star, including a number of favorable "soft" and "dynamic" effects associated with the program.[8,9] After reviewing the evidence on Green Lights and Energy Star, Howarth et al. (2000) concluded that "voluntary agreements between government agencies and private sector firms can ... lead to improvements in both technical efficiency of energy use and the economic efficiency of resource allocation." Unfortunately, none of these studies was able to distinguish between the improvements attributable to the voluntary programs and those changes that likely would have taken place even without the programs.

The empirical evidence is more extensive, though still mixed, when we look at programs focused explicitly on environmental performance as opposed to technology adoption, particularly with regard to toxics where there has been extensive analysis using TRI data. What is probably the gold standard in the field is an in-depth analysis of the 33/50 program by Khanna and Damon (1999), who jointly modeled the decision to participate in the program as well as the actual outcomes. They first recognize that a firm's decision about the quantity of covered releases to emit will likely depend on both its participation in 33/50 and such factors as stakeholder pressure, output levels, and others. They then allow for the participation decision to both depend on these same variables and to be correlated with the volume of releases. Using publicly available firm-level data, they found a statistically significant impact of the program on toxic releases, as well as on firms' return on investment and long-run profitability. Khanna and Damon hypothesize that the incentives for participation arise from three sources: program features, the threat of mandatory environmental regulations, and firm-specific characteristics.[10]

Focusing on the period 1988–1995, Sam and Innes (2005) also found that participation in 33/50 lowered releases of the covered chemicals, particularly in 1992. Further, they found that participation in 33/50 was associated with a significant decline in EPA inspection rates for the years 1993–1995. A study by Gamper-Rabindran (2005) found that while the effects varied by industry, in the case of the largest participating industry, namely the chemical industry, the positive results that 33/50 reduced toxic releases (reported by Khanna and Damon [1999]) are actually reversed when the analysis excludes two ozone-depleting chemicals whose phase-out was mandated by the Clean Air Act. Khanna has authored a new case study in this volume that reviews and updates the extensive literature on 33/50.

King and Lenox (2002) analyzed the environmental impact of firms participating in Responsible Care, an industry-sponsored effort to cut toxic releases distinct from the government-sponsored 33/50. Using pooled and panel data for the period 1991–1996, they find that participants were reducing their releases more slowly than nonparticipants. Their fixed-effect model shows that Responsible Care had an insignificant effect on environmental performance. That is, despite the improved performance of the chemical industry over the studied period, the rate of improvement was not greater than in pre-program years and, most surprisingly, it was slower for participants in Responsible Care than for non-members.

Moving Beyond Toxics, a paper by Dasgupta, Hettige, and Wheeler (2000) focused on the adoption by Mexican firms of ISO 14001 management practices. They found a significant improvement in the (self-reported) compliance status of participating firms. They also found that explicit environmental training programs for nonenvironmental workers led to an improvement in the compliance status of the firms.

Turning to energy and climate change, an analysis of the U.S. Department of Energy's Climate Challenge Program on CO_2 emissions focused on the largest 50 electric utilities east of the Rocky Mountains from 1995–1997 (Welch, Mazur, and Bretschneider 2000). Despite a number of intriguing results about the motivation of firms to participate in Climate Challenge, the authors find that adoption of the program seems to have no effect on emissions. In fact, those firms predicted to volunteer higher reduction levels were found to reduce their CO_2 emissions less. The authors hypothesize that the poor program performance is associated with the lack of at least a tacit regulatory stick of the type present in 33/50.

Overall, the literature is characterized by a paucity of empirical studies on the actual environmental performance of voluntary programs and, equally important, an almost exclusive focus on toxics as opposed to energy- or climate-related programs. As is well known, energy issues differ from toxics in many ways, including the extent to which financial incentives are already in place to reduce emissions. That is, market forces already encourage conservation and energy efficiency, whereas no such forces exist to reduce toxic emissions. Thus, the potential for voluntary programs to achieve reductions in energy-related carbon dioxide emissions may be more limited than the potential associated with toxics. A key motivation for this volume is to increase the attention paid to the rigorous study of program results and to emphasize rapidly growing interest in energy- and greenhouse-gas-related programs.

Programs Studied in This Volume

The seven case studies presented in this volume cut across multiple sectors, industries, and countries and represent a range of different approaches to achieving voluntary cooperation of participating firms and, in one case, households. As noted, only one focuses on toxic releases, while the other six involve energy use or greenhouse gas emissions. The programs entail a mixture of unilateral agreements, public voluntary programs, and negotiated agreements. The selected programs, case study authors, and other descriptive information are shown in Table 1-1.

No claim is made that these cases are representative of the thousands of voluntary programs in operation in the United States, Europe, or Japan. Rather, the cases were chosen somewhat opportunistically, largely because of the existence of the type of information needed to conduct an in-depth study with a focus on environmental performance. The selected programs are of sufficiently high profile to warrant detailed studies and they were established early enough—mostly in the 1990s—so there is now an actual record of their performance. Because our focus is on comparing the environmental results of participants in voluntary programs to a credible baseline, most of the selected programs are ones for which either a representative sample of nonparticipants or a historical/negotiated baseline is available. Arguably, these selection criteria omit newer, potentially improved programs without a track record, as well as those for which a comprehensive sample of nonparticipants or baseline is not readily available. These limitations must be balanced against the consequent depth and rigor of the analysis possible with the programs we do examine. Overall, we believe the insights gained from this approach for evaluating the environmental performance of voluntary programs outweigh the limitations.

TABLE 1-1. Selected Characteristics of Case Studies

Program	Author(s)	Years of operation	Energy, CO_2 (GHGs), or toxics	Industry or household	Program type
33/50 (U.S.)	Khanna	1991–1996	Toxics	Industry	Public voluntary program
Japanese Keidanren	Wakabayashi and Sugiyama	1997–	CO_2	Industry	Negotiated agreement
UK Climate Change Agreements	Glachant and de Muizon	2001–	CO_2	Industry	Public voluntary program
Danish Energy Efficiency Agreements	Krarup and Millock	1996–	CO_2	Industry	Negotiated agreement
German cement industry	Böhringer and Frondel	1995	CO_2	Industry	Unilateral agreement
Climate Wise (U.S.)	Morgenstern, Pizer, and Shih	1993–2000	GHGs	Industry	Public voluntary program
California demand-side management	Sanstad	Early–mid 1990s	Energy	Household	Public voluntary program

Note: GHGs = greenhouse gases.

The chapter authors themselves represent a diversity of expertise on voluntary programs, each with considerable prior knowledge of and familiarity with the subject programs. As part of the process for developing and evaluating the contributed chapters for this volume, an authors' workshop was held at Resources for the Future in December 2005 to discuss the cross-cutting methodologies and initial results of the individual studies. In the course of the workshop, outside reviewers provided comments on each of the chapters.[11] As editors, our goal has been to leverage the authors' previous work and to press them to develop practical and accessible information for the policy community on the environmental performance of these programs.

Questions for the Case Study Authors

The assignment for each author or team was to consider three broad issues related to the individual voluntary programs: 1) the context; 2) the design; and 3) the performance of the programs.

Program Context: We asked authors to begin by describing the circumstances surrounding the voluntary program. This question can be broken down into several pieces. First, what events have taken place, or are expected to take place, that might influence either program participation or outcome? For example, rising energy prices that coincide with the instigation of an energy-efficiency program might affect both participation in the program and energy use directly. Or rising public concern about toxic chemicals might spark interest both in a voluntary program and in action to thwart that concern. Second, does creation of the voluntary program signal anything about the future? Many people in the United States, for example, might have viewed President George W. Bush's announcement of new, voluntary climate initiatives as a signal that any mandatory climate policies remain a ways off. Similarly, negotiated agreements in Europe might be viewed as forestalling regulations in the sectors where they exist.

Understanding program context is important to comparing both participation and outcomes across countries. Otherwise identical programs where one carries the threat of further action or where one is accompanied by changes in energy policy or prices may see dramatically different outcomes. While some contextual features clearly are associated with national differences, it is important to draw them out in the comparison. In the United Kingdom and Danish climate change programs, for example, participants were offered an 80–100% reduction in the newly adopted energy tax in exchange for a quantified commitment to reduce their energy or CO_2 emissions.

Program Design: We next asked authors to describe the goal of the program and what participation requires. Programs can be categorized as to whether participation is based on qualitative or quantitative commitments. Qualitative commitments might involve reporting emissions or abatement activities undertaken or they could require use of particular technologies, such as improved insulation or efficient light bulbs. Quantitative commitments would be to reduce emissions or energy use in either absolute or relative (to output) terms.

We are particularly interested in how program design features affect participation and outcomes. Policymakers may have little influence over national circumstances but certainly can affect program design. Some of the most important lessons we hope to draw are how different features increase or decrease program effectiveness.

Program Outcomes: How has the program been evaluated, and what were the results of that evaluation? There are a variety of ways that programs can be evaluated, ranging from self-reported results, or even casual observations, by participants or program administrators, to careful studies that attempt to simulate the effect of a randomized experiment. In some cases, there may be quantitative reports on the outcome of interest—energy use or emissions—and in some cases there may be only qualitative information about participation, "soft effects," or long-term assessments. While our choice of programs and authors is based in part on their ability to provide an analysis based on their existing work, several authors have conducted entirely new quantitative analyses for this volume.

How should outcomes be measured? Absent a randomized experiment, we are forced into making assumptions about the appropriate baseline from which to measure progress for program participants. Perhaps there are historic outcomes or other information that can be used to forecast a business-as-usual baseline for comparison to results after the program's inception. Perhaps data exists on similar groups of participants and nonparticipants. Finally, there may be reports of actions and consequences related to participation, with the assumption that the actions would not have happened absent the program. All these non-experimental approaches have some flaws and potentially give rise to bias in the estimated program benefits. We have encouraged authors to discuss these biases in their analyses.

Variation in baselines, as well as outcome metrics, could pose problems for our comparison exercise but also presents opportunities. It is important to understand why some programs appear to have been more successfully evaluated than others and whether there were program design details that might make future evaluation easier. It also is valuable to consider other quantitative indicators, such as participation levels, as well as qualitative results reported by stakeholders, such as shaping attitudes and improving management. We anticipate that by putting policy evaluations across programs and countries side-by-side, we will see patterns of effectiveness even when outcome measures are not directly comparable. The idea that the collective evaluation of multiple programs will be more compelling than the sum of the individual evaluations is, in part, what motivates this exercise.

Advice to the Reader and Practitioner

Despite the preceding caveats, readers are encouraged to focus, as the authors do, on the program outcomes. Do these programs generate significant environmental improvements without the burdens associated with mandates? How large are the gains? Are they greater than those likely to have been achieved without the program in place? Is it realistic to expect the programs to substitute for formal regulation over the long term or are they just a way to start the process in the absence of legal or political mandates to go further?

Another issue to think about is the incentives for firms to join voluntary programs in the first instance and, particularly, the effect of the different incentives on the likelihood of joining or actually improving environmental performance. In fact, the incentives in place in the seven case studies are quite disparate. Two of the programs reward participants with specific rebates from major new energy or carbon taxes adopted to address climate change. In several other programs there is a clear expectation of mandatory requirements if firms fail to join and, most importantly, reduce their emissions. In others, the expectations are less clear. The key question is how or to what extent do these incentives affect environmental results?

A related issue is the selection of targets for the programs, as highlighted by the taxonomy described earlier. In some cases the targets were selected unilaterally by industry. In others, the government set ground rules which firms or households agreed to. In yet others, there was a two-way dialogue between participants and government to arrive at specific terms. Does government involvement make a difference in the stringency of the targets established? Is there evidence of regulatory capture, wherein the targets chosen may represent little more than business as usual?

Finally, there are some interesting methodological issues running through the seven cases—issues that affect the robustness of the evaluations themselves and, quite possibly, might have application in the design of future programs. As noted, 33/50 is unique in that the reporting system, the TRI, applies equally to participants and nonparticipants alike. None of the other programs studied has such a built-in reporting scheme. Yet, all the other authors have endeavored to address the same basic issues about program performance, with some using a control group and others relying on a forecast baseline. The reader should consider the different methods and assess their relative usefulness for evaluation.

We return to all of these issues and consider how they cut across the different cases in the final chapter of the volume.

References

Alberini, Anna, and Kathleen Segerson. 2002. Assessing Voluntary Programs to Improve Environmental Quality. *Environment and Resource Economics* 22(1–2): 157–184.

Arora, Seema, and Timothy Cason. 1995. An Experiment in Voluntary Environmental Regulation: Participation in EPA's 33/50 Program. *Journal of Environmental Economics and Management* 28: 271–286.

———. 1996. Why do Firms Volunteer to Exceed Environmental Regulations? Understanding Participation in EPA's 33/50 Program. *Land Economics* 72: 413–432.

Baranzini, Andrea, and Phillipe Thalmann. 2004. *Voluntary Approaches in Climate Policy.* Cheltenham, UK: Edward Elgar.

Celdren, A., H. Clark, J. Hecht, E. Kanamaru, P. Orantes, and M. Santaello Garguno. 1996. The Participation Decision in a Voluntary Pollution Prevention Program: The U.S. EPA 33/50 Program as a Case Study. In *Developing the Next Generation of the U.S. EPA's 33/50 Program: A Pollution Prevention Research Project*, edited by H. Clark. Durham, NC: Nicholas School of the Environment, Duke University.

DeCanio, S. J. 1998. The Efficiency Paradox: Bureaucratic and Organizational Barriers to Profitable Energy-Saving Investments. *Energy Policy* 26: 441–454.

DeCanio, S. J., and William E. Watkins. 1998. Investment in Energy Efficiency: Do the Characteristics of Firms Matter? *Review of Economics and Statistics* 80: 95–107.

Dowd, Jeff, Kenneth Friedman, and Gale A. Boyd. 2001. How Well Do Voluntary Agreements and Programs Perform at Improving Industrial Energy Efficiency? Washington, DC: Office of Policy, U.S. Department of Energy.

EC (Commission of the European Communities). 1996. *On Environmental Agreements*, Communication from the Commission to the Council and the European Parliament. Brussels: EC.

EEA (European Environment Agency). 1997. Environmental Agreements: Environmental Effectiveness. *Environmental Issues Series No. 3* (Vols. 1 and 2). Copenhagen: EEA.

Gardiner, David. 2002. The Effectiveness of Voluntary Programs and Next Steps in Climate Change Policy. In *U.S. Policy on Climate Change: What Next?* edited by John A. Riggs. Washington, DC: The Aspen Institute.

Howarth, R., B. Haddad, and B. Paton. 2000. The Economics of Energy Efficiency: Insights from Voluntary Programs. *Energy Policy* 28: 477–486.

Khanna, Madhu. 2001. Non-Mandatory Approaches to Environmental Protection. *Journal of Economic Surveys* 15(3): 291–324.

Khanna, Madhu, and Lisa A. Damon. 1999. EPA's Voluntary 33/50 Program: Impact on Toxic Releases and Economic Performance of Firms. *Journal of Environmental Economics and Management* 37(1): 1–25.

Khanna, Madhu, and D. Ramirez. 2004. Effectiveness of Voluntary Approaches: Implications for Climate Change Mitigation. In *Voluntary Approaches in Climate Policy*, edited by A. Baranzini and P. Thalmann. Cheltenham, UK: Edward Elgar.

King, A. A., and M. J. Lenox. 2002. Does Membership Have Its Privileges? Working paper. New York: Stern School of Business, New York University.

Lyon, T., and J. Maxwell. 2002. Voluntary Approaches to Environmental Regulation: A Survey. In *Economic Institutions and Environmental Policy*, edited by M. Franzini and A. Nicita. Hampshire, UK: Ashgate Publishing.

Sam, A. G., and R. Innes. 2005. Voluntary Pollution Reductions and the Enforcement of Environmental Law: An Empirical Study of the 33/50 Program, Manuscript, Department of Agricultural and Resource Economics, University of Arizona, Tucson.

Segerson, Kathleen, and N. Li. 1999. Voluntary Approaches to Environmental Protection. In *The International Yearbook of Environmental and Resource Economics 1999/2000*, edited by H. Folmer and T. Tietenberg. Cheltenham, UK: Edward Elgar.

Starik, M., and A. Marcus (eds.). 2002. Management of Organizations in the Nature(al) Environment, Special Research Forum. *Academy of Management Journal* 43(4): 539–546.

Tsutsumi, R. 1999. The Nature of the Voluntary Agreement in Japan. Paper presented at the European Research Network on Voluntary Approaches (CAVA) Workshop. May 1999, Copenhagen, Denmark.

U.S. EPA (U. S. Environmental Protection Agency). 2005. *Everyday Choices: Opportunities for Environmental Stewardship*. Technical Report by the EPA Environmental Stewardship Staff Committee for the EPA Innovation Council. Washington, DC: U.S. Environmental Protection Agency.

Welch, E., A. Mazur, and S. Bretschneider. 2000. Voluntary Behavior by Electric Utilities: Levels of Adoption and Contribution of the Climate Challenge Program to the Reduction of Carbon Dioxide. *Journal of Policy Analysis and Management* 19(3): 407–425.

Notes

1. This point has been emphasized by the European Environment Agency (1997).

2. For a discussion of the limits of voluntary programs, see Gardiner (2002).

3. In the case of the DSM programs, households are the target of participation rather than firms, and utilities play the role of the government agency, establishing the terms of the program.

4. For reviews of this literature see Khanna (2001) and Lyon and Maxwell (2002); see also Arora and Cason (1995, 1996); Celdren et al. (1996); Khanna and Damon (1999); and OECD.

5. For reviews of this literature see Khanna (2001), Lyon and Maxwell (2002), Alberini and Segerson (2002), Khanna and Ramirez (2004), and Baranzini and Thalmann (2004).

6. Guidelines issued by the European Commission in 1996 were designed to address evaluation issues, in part, by requiring greater quantification of objectives, better monitoring, and more transparency in reporting. To our knowledge, no recent assessment of these European programs has been conducted since the guidelines became effective.

7. A substantial literature exists on the characteristics of firms that join voluntary programs of all types. For example, in addition to Green Lights, studies have been done on the characteristics of firms joining 33/50, Climate Challenge, Waste Wise, as well as for adopting ISO 14001. See Khanna (2001) for discussion.

8. For example, citing work by Laitner (2001), they note "increased awareness of the benefits of energy-efficiency equipment." They also note that "80 percent of consumers surveyed were familiar with the logo; and more than 40 percent used the logo in purchase decisions."

9. See also Oak Ridge (2000) and Laitner and Sullivan (2001).

10. Khanna and Damon (1999) consider the threat of mandatory penalties by comparing firms currently listed as responsible parties for a larger number of Superfund sites to those with a smaller number of sites—arguing that the former are more likely to be aware of the liability costs of continuing to generate their past levels of toxic pollution. Similarly, they model a series of firm-specific characteristics such as innovativeness, age of existing equipment, membership in industry trade associations, and volume of toxic releases that might be expected to influence the costs and benefits of participation and thus the participation decision.

11. The following individuals served as peer reviewers: Tom Lyons, Terry Dinan, Bob Shackleton, Joe Aldy, Janet Peace, Andreas Lange, and Kathleen Hogan.

2

The U.S. 33/50 Voluntary Program
Its Design and Effectiveness

Madhu Khanna

Environmental regulation in the United States primarily has relied on mandatory command-and-control regulations that prescribe quantity limits on emissions of pollutants or use of specific abatement technologies. This has led to media-specific, end-of-pipe pollution control, encouraged cross-media substitution of pollutants, and imposed steeply rising costs of abatement on firms and regulators. The passage of the Emergency Planning and Community Right to Know Act of 1986, which required manufacturing facilities to report their annual on-site releases and off-site transfers of specified toxic chemicals to the Toxics Release Inventory (TRI), raised awareness among regulators and the public about the magnitude of the toxic pollution problem. The task of designing command-and-control methods for hundreds of unregulated toxic pollutants, however, would have been administratively difficult, slow, costly, and possibly beyond the budgets of regulatory agencies. Regulatory agencies needed quick, cost-effective, and less adversarial approaches to reducing toxic pollution. Encouraging voluntary action by firms to reduce their toxic releases was one such approach. It was also the approach chosen for the implementation of the Pollution Prevention Act of 1990, which established a hierarchy for waste management, with prevention and reduction given top priority, followed by recycling/reuse, treatment, and disposal.

In 1991, the U.S. Environmental Protection Agency (EPA) launched its first voluntary program, called the 33/50 program. The program sought to reduce emissions of 17 high-priority toxic chemicals reported to the TRI through voluntary action by firms. It got its name from its two numeric goals of reducing the level of on-site releases and off-site transfers of the 17 designated chemicals (referred to as 33/50 releases, hereafter) by 33% by the end of 1992 and by 50% by 1995 relative to their 1988 levels.

In contrast to the command-and-control approach used by EPA in the past, the 33/50 program sought to encourage pollution prevention rather than end-of-pipe treatment technologies by establishing a voluntary partnership among governments, communities, and industries (U.S. EPA 1992). It was an innovative attempt to demonstrate that voluntary partnerships could augment the agency's traditional command-and-control approach by bringing about targeted reductions more quickly than through regulations alone. Chris Tirpak of EPA's 33/50 program staff remarked, "It's a paradigm shift. We are moving from confrontation on environmental problems to collaboration on environmental solutions. . . . This is a bridge to a new way of doing business. It gives industry an incentive to reduce chemical releases and transfers."[1]

EPA believed that the realization by firms that a reduction in their toxic releases would not only benefit the environment but also their bottom line would make firms willing to engage in this partnership with the government.[2] EPA also hoped to instill a pollution-prevention ethic among firms and promote efforts to make continuous environmental improvements through pollution prevention (U.S. EPA 1992). Underlying this belief was the premise that pollution represents wasteful use of resources and that pollution prevention pays. By preventing waste voluntarily, firms may increase efficiency in the use of polluting inputs, increase demand from environmentally conscious consumers, discourage stakeholder activism, and attract better workers (Hart 1995; Porter and van der Linde 1995; Reinhardt 1999). Since participation in the program was expected to benefit firms financially, no monetary incentives were offered to induce participation.

In 1988, there were 6,000 firms that reported to the TRI that they emitted 33/50 releases. The aggregate amount of 33/50 releases was 1.5 billion pounds. By 1995, these releases had declined by 824 million pounds (55%). The trend continued in 1996, with 33/50 releases declining by 896 million pounds (60%) over the 1988 level. The 33/50 program raised several key issues that are addressed in this case study. What factors led firms to agree to voluntarily participate in the 33/50 program? To what extent were the achieved reductions in 33/50 releases due to the program and would not have been achieved otherwise? How did the design of the 33/50 program influence its performance? Did program participation improve the financial performance of firms as expected by participants and EPA? In other words, was it rational for profit-maximizing firms to participate in the program in the absence of any government financial incentives for doing so?

This chapter traces the history behind the development of the 33/50 program and provides a context for the choice of pollutants targeted by the program. It describes the implementation of the 33/50 program and its design, such as the incentives for participation and the methods for evaluating the performance of the program. The accomplishments of the program from the viewpoint of those directly involved—EPA and the firms—are discussed as well as the motivations for firms to participate in the 33/50 program. The chapter also provides an analysis of program performance from the viewpoint of academics, government, and nongovernmental organizations.

Background

Since the early 1970s, EPA's structure and approach toward environmental regulation have been driven by media-specific or pollutant-specific statutes that authorize the agency's programs. This approach made it difficult for EPA to base its priorities on an assessment of risk across all environmental problems. The need to address this issue was recognized by the mid-1980s when the then-EPA administrator called on the agency to allocate its resources so that they account for the relative risks posed by environmental problems; recognize that pollution-control efforts in one medium can cause pollution problems in another; and achieve measurable environmental results. The toxic gas leak at the Union Carbide plant in Bhopal, India, in 1984 focused public attention on the potential for chemical accidents and highlighted the lack of understanding about releases and transfers of toxic substances in the United States. It led to the passage of the Emergency Planning and Community Right to Know Act in 1986, which imposes annual reporting requirements on firms releasing designated toxics into the environment. It also requires that EPA make available to the public facility-specific data reported to the TRI. EPA began collecting these data in 1987 and making them available to the public with a lag of two years.

Although the TRI did not impose any requirements for reducing toxic releases, the disclosure of previously unanticipated amounts of toxic releases by firms was expected to raise public awareness and concern about the toxic waste problem and create pressure on firms to reduce these releases. The TRI also was expected to make firms realize the magnitude of their own emissions. In anticipation of negative publicity, the CEO of Monsanto promised on June 30, 1988 (the day before the first TRI reports were due at EPA) that by 1992 the firm would reduce its worldwide releases of toxic air emissions by 90% from 1987 levels (Kirschner 1995). Other firms such as Baxter and Lilly also faced considerable adverse publicity in 1989. Consequently, these and other such firms began their efforts to reduce toxic releases as early as 1989, even prior to the 33/50 program.

In September 1990, a report by the Science Advisory Board (SAB) of EPA highlighted the importance of addressing human health risks due to direct public exposure to known toxic agents (U.S. EPA 1990). In response to this report, then-EPA director William K. Reilly proposed an ambitious strategy for reduction of 15 or so toxics "by one-third by the end of Fiscal Year 1992 and by more than half by 1995, through the most cost-effective methods possible."[3]

This later developed into the 33/50 program. The SAB report also emphasized reduction of chemical wastes at source as the preferred option for reducing risk, a multi-media approach to pollution control, and reliance on tools such as public education and information provision, technical assistance, and market incentives instead of command-and-control approaches. This, together with the passage of the Pollution Prevention Act in 1990, strongly influenced the design of the 33/50 program (INFORM 1995).

Pollution prevention began to receive significant attention from federal and state governments, large companies, and the industrial trade press in the 1990s. Many large corporations, including 3M, Monsanto, DuPont, Polaroid, and AT&T, established internal programs for preventing pollution, such as 3M's Pollution Prevention Pays. By 1989, 26 states had established some form of pollution-prevention, waste-reduction, or toxics use-reduction program. These state programs lacked any specific, quantifiable measures and systematic evaluation protocols for quantifying progress in pollution prevention. Their legislation did not provide any guidance for targeting specific pollutants. The states did not establish any priorities for the chemicals to be reduced from the more than 300 chemicals listed in the TRI. The 33/50 program represented a first step toward setting environmental priorities among pollutants based on relative risk and setting quantifiable targets for pollution reduction.

Each of EPA's media-specific programs was asked to identify 15 or so toxic releases associated with high-volume industrial chemicals that posed the most serious environmental and health risks and whose release potentially could be reduced through pollution prevention. This list led to the identification of the 17 high-priority chemicals targeted by the 33/50 program. These chemicals are listed in Table 2-1.

These chemicals were used in large quantities by U.S. manufacturing facilities, they had a high ratio of releases to usage (suggesting inefficiencies in the production process), and they could be reduced through pollution prevention (Prakash 2000). In 1988, these 33/50 releases accounted for 22% of the total quantity of toxic releases reported to the TRI and were emitted by 57% of the more than 21,000 TRI facilities (U.S. EPA 1992).

In an effort to provide results to the public more expeditiously than would have been possible with mandatory regulations, EPA decided to implement a voluntary program seeking cooperative action from firms for reducing releases of these chemicals. EPA administrator Reilly, however, stated that this did not imply that conventional approaches to environmental problems not cited as high risk would be abandoned. EPA also stressed that the 33/50 program was not a substitute for conventional regulations and that EPA remained committed to continued, intensified enforcement of environmental laws already on the books. But the program represented a movement toward targeting the agency's limited resources to address high-risk environmental problems through voluntary actions.[4]

TABLE 2-1. 33/50 Chemicals and Their Releases in 1988

Names of 33/50 chemicals	Releases and transfers in 1988 (million pounds)	Percentage of total 33/50 releases in 1988
Benzene	36.8	2.5
Carbon tetrachloride	5.3	0.4
Chloroform	29.8	2.0
Dichloromethane	155.4	10.4
Methyl ethyl ketone	171.8	11.5
Methyl isobutyl ketone	45.0	3.0
Tetrachloroethylene	42.5	2.8
Toluene	367.5	24.6
1,1,1-Trichloroethane (TCA)	200.9	13.4
Trichloroethylene	62.6	4.2
Xylenes	212.9	14.2
Cadmium and cadmium compounds	1.8	0.1
Chromium and chromium compounds	71.2	4.8
Cyanide compounds	12.0	0.8
Lead and lead compounds	60.9	4.1
Mercury and mercury compounds	0.3	0.0
Nickel and nickel compounds	19.6	1.3
Total	1,496.3	100

Source: U.S. EPA (1999).

To induce such voluntary actions by firms, EPA conveyed to them that the 33/50 program would be a prototype for new laws if "voluntary" compliance from industry was not forthcoming. A letter written by EPA Administrator Reilly to the CEO of Baxter noted that (see Prakash 2000): "The American public has made clear that they expect nothing less than dramatic reductions in toxic chemical releases. The challenge before EPA and industrial leaders is this: how do we bring such reductions about? One way is by the conventional command-and-control option that has been the agency's mainstay for the past twenty years. But I believe there is another, more fruitful path that we follow which is faster, and without the detailed direction which is likely to be demanded by the public if voluntary efforts are not fruitful."

Regulations were already imminent for some types of 33/50 releases. Air emissions of the 17 toxic pollutants (which were included in the 189 air pollutants classified as hazardous air pollutants) had been targeted for quantity limits under the provisions of the 1990 Clean Air Act. EPA was expected to set Maximum Available Control Technology (MACT) standards for these air pollutants by the year 2000. These MACT standards were to be based on emissions levels already being achieved by the best-performing similar facilities. By participating in the 33/50 program and reducing releases, firms could establish themselves as environmental leaders in the industry and shape the MACT standards to be set by EPA. They could also use the reductions achieved under the 33/50 program to qualify for the early reduction incentive provided by the Clean Air Act Amendments (GAO 1994; U.S. EPA 2000).[5]

Additionally, two of the organic substances targeted by the 33/50 program were ozone-depleting chemicals, carbon tetrachloride and 1,1,1-trichloroethane (TCA), that were targeted for phasing out by January 1, 1996, by the Montreal Protocol. In implementing this agreement, the United States had banned production of carbon tetrachloride and TCA by that deadline. Carbon tetrachloride accounted for a relatively small portion (5.3 million pounds) of total 33/50 releases in 1988, but with 201 million pounds of releases and transfers in 1988, TCA accounted for a much larger portion of 33/50 releases (see Table 2-1).

EPA believed that companies would be willing to participate in the program to achieve cost-savings through increased efficiency resulting from efforts to reduce 33/50 releases and to receive public recognition through EPA. The minimum prerequisites for joining the program and flexibility in determining the amount and the methods for pollution reduction were also expected to reduce the disincentives for participation.

Many firms were favorably inclined toward the 33/50 program. After release of the TRI in 1989, hundreds of articles singling out the top polluters—the "dirty dozens" in states and counties—had appeared in local and national media (see for example, *New York Times* 1991). The 33/50 program provided these large polluters with a means of having their progress toward toxic pollution reduction recognized through a formal EPA program that would receive broad public attention. These firms also saw the possibility of a win-win situation whereby they could reduce emissions and increase profits by preventing waste generation, forestalling tougher regulations and gaining stakeholder goodwill. They recognized that the alternative of command-and-control regulations to control these releases would not give them much autonomy on issues such as chemicals to reduce, technology to use, and the time frame for reduction.

In 1992, Dow's president and CEO Frank P. Popoff (1992) stated, "I commend the 33/50 program.... If we wait for more regulation, which is sure to come if we don't make tangible improvements, there will be only 'end-of-pipe' type treatments installed as we react to legislation and regulatory pressure—hardly a mandate for pollution prevention.... If pollution prevention is really going to catch on, it will require a change of approach by all of us, by all the major players starting right now."

A voluntary program has the potential to allow firms to be more proactive about improving their environmental performance and can provide incentives for firms to go beyond compliance. "Voluntary programs have created a new mind-set," said Paul V. Tebo, vice president for environment, health, and safety at DuPont. "The direction is terrific and we're connecting with EPA. Today, most companies are not considering regulations the issue. We began to challenge our corporation to move ahead with a whole range of commitments. Before that, the primary driver was EPA. [Voluntary efforts] have created a very healthy competition, a huge change," he noted (Kirschner 1995).

There are several aspects that need to be considered in designing a voluntary program: setting the goals of the program; incentives to be provided for participation; conditions for participation; and methods for evaluating program performance. The design of the 33/50 program is discussed below.

Program Implementation

The baseline level and goals for 33/50 releases were set with reference to the level in 1988. At the time the 33/50 program was formulated, 1988 was the most recent year for which TRI data were available. EPA also chose 1988 as the baseline year with the intention of permitting companies to take credit for programs that were already in place and not to penalize them for taking steps to cut pollution prior to 1991 (BNA 1994). The program sought to reduce only on-site releases and off-site transfers for treatment and disposal of the 33/50 chemicals; facilities were required to report only these data to the TRI in 1988. On-site releases included air emissions, surface water discharges, underground injection, and on-site releases to land. Off-site transfers included wastes transferred for treatment and disposal by private facilities or by publicly owned treatment works. The effects of the program on off-site transfers to recycling and energy recovery and on on-site waste management were not considered since firms were required to report these to the TRI only from 1991 onwards.

Participation in the program was voluntary. EPA used the 1988 TRI data to identify more then 12,000 facilities owned by more than 6,000 parent companies that reported releases and transfers of one or more of the 33/50 chemicals. To encourage participation in the program, EPA sent letters to the CEOs of the parent companies of the facilities that were emitting 33/50 releases. By contacting the CEOs of the parent companies, EPA sought to encourage a pollution prevention philosophy among the highest echelons of corporate America and to encourage participation from every facility within the parent company. In January 1991, the EPA administrator sent letters to the CEOs of the "Top 600" companies inviting them to participate in the 33/50 program. The facilities of these "first-round" companies accounted for 66% of the 33/50 releases in 1988. A series of meetings with top executives from different industrial manufacturing sectors, trade association leaders, and company managers were held to discuss the program's implementation. The 33/50 program was publicly announced in February 1991.

In July 1991, letters of invitation were sent to 5,400 "second-round" companies. To encourage participation by the smaller, second-round firms, EPA tried new outreach approaches, such as calling facilities directly to discuss the benefits of the program and to identify and address any barriers that may prevent them from participating. A third round of invitations were sent after 1992 to 2,512 companies that started emitting 33/50 releases after 1988. A total of 8,512 companies (out of the 10,167 companies that emitted 33/50 releases between 1988 and 1995) had been invited to participate by January 1994. There were 2,612 companies that started emitting 33/50 releases after 1992 but were not contacted to join the program (U.S. EPA 1999).

The 33/50 program emphasized prevention of pollution upstream and wherever possible through reduction in the use of toxic chemicals rather than managing wastes after they have been generated. The program encouraged firms to develop less toxic substitutes, reformulate products, and redesign production processes to achieve source reduction rather than resort to end-of-the-pipe methods for pollution control. However, the program did not require use of any specific abatement methods. EPA also was flexible in terms of reduction goals chosen by a firm. The 33% and 50% reduction targets were national goals for aggregate 33/50 releases and not goals for individual firms or for individual chemicals. The goals of this program also were not media-specific—that is, focused specifically on emissions to air, land, or water; instead, firms could reduce releases to any media. Additionally, a firm could decide which of the 33/50 chemicals it wanted to reduce, as well as which of its facilities would participate in the program. EPA encouraged participants to set their own reduction goals, oriented to their own time frames. At a minimum, a letter of intent by a firm to participate in the program without specifying any numerical reduction goals was sufficient for it to be considered a participant. Participating companies could count reductions made in off-site transfers and on-site releases since 1988 against the reduction goals they set. EPA did not monitor emissions of the participating firms to ensure that the figures were accurate, but program participation did require the chief financial officer to sign a form stating the amount of chemicals released.[6] Firms submitting reduction commitments received a formal certificate of participation from EPA.

Incentives to participate in the program were primarily in the form of public recognition of companies that participated, as well as technical assistance to enable pollution prevention. The program publicized participation by firms through its annual Progress Update Reports, through press releases, and through an awards program to recognize special efforts made by firms. It also did not penalize firms (through adverse publicity) if they did not meet their reduction commitments. Thus, the program sought to impose few costs on firms while providing firms with a great deal of flexibility in the extent of their participation. EPA stated that the program was enforcement neutral; participants would not receive preferential treatment of any kind in the form of relaxed regulatory oversight or enforcement of other EPA regulations nor would nonparticipants receive special scrutiny. Arora and Cason (1995) provide some anecdotal evidence, which supports this, although as discussed below, Sam and Innes (2005) find evidence that contradicts this.

To help companies implement pollution-prevention practices for reducing 33/50 releases, EPA facilitated information collection, coordination, and exchange. It conducted regional pollution-prevention workshops for industry to exchange information about pollution-prevention practices and established the Pollution Prevention Information Exchange System (a computer bulletin board) containing technical, policy, and financial information related to pollution prevention. It published bibliographic reports and training guides identifying pollution prevention documents, industry-specific guidance manuals, fact sheets, and videos. EPA also prepared and distributed company profiles describing reduction activities implemented by program participants.

EPA evaluated the 33/50 program by looking at the overall reduction in 33/50 releases and not by examining outcomes at the individual firm level. It compared the aggregate releases and transfers of these chemicals by participating and non-participating parent companies. EPA counted total releases and transfers from all facilities of participating companies regardless of whether there was a commitment covering each facility (even including those specifically excluded from the company's commitment). EPA stated that reductions achieved between 1988 and 1990 would be considered as contributing to the program's reduction goals (U.S. EPA 1999). These reductions, however, would not be viewed as resulting from the program, because companies were first informed about the program in February 1991. Measures of success identified by EPA were whether greater reductions in 33/50 releases occurred after 1991; whether a larger number of companies voluntarily used pollution-prevention methods to reduce 33/50 releases; and whether 33/50 releases fell faster than those of other TRI chemicals.

Although the program sought to encourage pollution prevention, EPA did not require firms to report on the quantity of releases prevented from entering the wastestream due to the adoption of pollution-prevention techniques. Instead, starting in 1991 under the Pollution Prevention Act of 1990, firms were required to report to the TRI the number and type of practices they adopted to prevent pollution of the TRI chemicals at source. They also began to report on the amount of releases recycled, treated, and disposed.

In order to measure the amount of 33/50 releases reduced through pollution prevention, EPA would have needed either: a) information on the quantities of chemicals that would have entered the wastestream if companies had not implemented source reduction activities (this could then be used together with information about the quantities of pollutants actually released to infer the amount of pollution prevented at source); or b) direct reporting by firms of the quantity of waste they reduced through pollution prevention and not simply on the number of pollution-prevention practices they adopted. The TRI did not require firms to provide either type of information.[7] Because the 33/50 pro-

TABLE 2-2. Participation in the 33/50 Program

| Parent companies contacted (actual number of eligible companies):[a] date | No. of participants (% of those contacted)[b] | Total 33/50 Releases in 1988 from: | | | Pledged reductions (% of quantifiable commitment) (million pounds) |
		All firms	Participants (million pounds)	Firms with quantifiable commitments	
First round "Top 600" (509): March 1991	328 (64)	993	809	697	327 (47)
Second round: 5,400 companies (4,534): July 1991	819 (18)	367	110	69	37 (54)
Third round: 2,512, July 1992– January 1994	140 (6)	45	14	12	6 (50)
Not contacted: 2,612	9 (0.3)	89	1	0	–
Total firms: 10,167	1,296 (17)	1,494	934	778	370 (48)

a. Eligible firms are those with facilities emitting 33/50 releases
b.Excludes firms with numerical goals not quantifiable to the 1988 baseline and firms with use reduction goals only; there were 318 such firms.
Source: U.S. EPA (1999).

gram relied solely on data reported to the TRI, EPA could not evaluate the success of the 33/50 program in preventing 33/50 releases. We turn next to the visible measures of program performance. Following that we discuss the views of external analysts who seek to identify the contribution of the program to the reduction in 33/50 releases.

Program Accomplishments: The View of EPA and Firms

Participation

Of the "Top 600" firms contacted to participate in the program, 64% (328 firms) agreed to participate (Table 2-2). Of the firms invited to participate in the second round, 18% (819 firms) agreed to participate. Among the firms invited after 1992, less than 10% agreed to participate; while among the remaining firms not contacted, only 9 firms participated in the program. A total of 1,294 firms (17% of those eligible) agreed to participate in the program. Among these, the firms that joined the program in 1991 and 1992 accounted for 61% of the total 33/50 releases in 1988. There were 779 firms that made numerical pledges for reducing their 33/50 releases relative to 1988 levels. These pledges totaled 370 million pounds, representing a little less than half of their total releases and transfers of 778 million pounds. Not all pledges were for reduction beyond what the company had already accomplished at the time of the pledge. Other participant companies developed goals tied to changes in their production levels, chose baselines other than 1988 (because they did not meet the reporting threshold for the 33/50 releases in 1988), or set a reduction target for all TRI chemicals they were reporting without specifying goals for the 33/50 releases. Some companies agreed to participate without specifying numerical goals, while others pledged to reduce their use of the targeted chemicals (U.S. EPA 1999).

Companies gave a variety of reasons to EPA for their willingness to participate in the program. Some companies were already pursuing reduction efforts and welcomed the opportunity for formal recognition of their efforts. Other felt that goals of the 33/50 program and pollution prevention were consistent with total quality management principles, with goals of the Responsible Care program, and with responsibilities of firms to be good corporate citizens. Reasons for lack of participation were difficulties in predicting future releases due to uncertainties about company operations, lack of resources, and potential conflicts between efforts to reduce 33/50 releases and the goals of other environmental programs (U.S. EPA 1992).

Emissions Reduction

Pre-Program Period (1988–1990). TRI data showed that 33/50 releases fell by 15% from 1.5 billion pounds in 1988 to 1.27 billion pounds in 1990 (Table 2-3). During the same period, all other TRI releases fell by 12%, from 2.5 billion pounds to 2.2 billion pounds.[8] The companies that chose to participate in the program had achieved much greater reduction (21%) than companies that did not participate (7%) by 1990.

Program Years (1991–1995). By 1992, the 33/50 releases had fallen by 40% relative to the 1988 baseline and by 29% relative to the 1990 level. The program thus exceeded its interim goal of a 33% reduction relative to 1988 levels. Program participants reduced their 33/50 releases by 15% between 1991 and 1992; nearly twice the rate of reduction achieved by non-participating companies. Participating companies accounted for 80% of the reduction in 33/50 releases between 1991 and 1992. The 33/50 releases declined at four times the rate reported for all other TRI releases between 1991 and 1992 (10.4% vs. 2.6%) (U.S. EPA 1994).

By 1995 when the program ended, 33/50 releases had fallen by 824 million pounds (55%) relative to 1988. Of this reduction, 72% occurred after 1991. Between 1991 and 1995, participants reduced 33/50 releases by 55%, while nonparticipants reduced them by 36%. Of the 779 participants that had made quantifiable commitments, 593 firms met or exceeded their commitments. They reduced their releases by 73% relative to 1988 levels and 65% relative to 1990 levels. The remaining 186 firms made pledges for reductions they had already achieved since 1988. They reduced their emissions by 15% between

TABLE 2-3. 33/50 Releases by Program Participants and Nonparticipants

Company type and number	1988 (million pounds)	1990 (million pounds) (% change 1988–1990)	1995 (million pounds) (% change 1991–1995) (% change 1988–1995)
33/50 releases by:			
Participants that met or exceeded pledge (593)	675	526 (-22%)	182 (65%) (-73%)
Participants that did not meet pledge (186)	104	88 (-14.8%)	90 (1%) (-13.8%)
All participants (1,294)	935	743 (-20.5%)	338 (-55%) (-64%)
Nonparticipants (8,873)	562	522 (-7.1%)	335 (-36%) (-40%)
Total (10,167)	1,497	1,265 (-15.4%)	672 (-47%) (-55%)
Releases and transfers of other TRI chemicals	2,524	2,163 (-14.3%)	1,617 (-25%) (-36%)

Note: TRI = Toxics Release Inventory.
Source: U.S. EPA (1999).

1988 and 1990 and increased their emissions by 1% between 1990 and 1995. This suggests that there was considerable heterogeneity in the responses of firms to the program and in their ability to reduce emissions beyond levels achieved by 1991. The level of aggregate 33/50 releases continued to fall after 1995, and by 1996 they had fallen by another 11% of the 1995 levels. Releases by participants fell by 8%, whereas those by nonparticipants fell by 11% relative to 1995 (U.S. EPA 1999).

Participating firms had higher 33/50 releases in 1988 and continued to remain larger emitters in 1996 even though they made larger reductions. In 1988, average releases of a participating company were 0.72 million pounds, while those of a non-participating company were 0.06 million pounds. While the former reduced their releases to 0.23 million pounds in 1996, the latter reduced them to 0.03 million pounds in 1996. Participating companies also had achieved much larger reductions prior to the start of the program in 1991. They had reduced their emissions by 21% between 1988 and 1990, while nonparticipants had reduced their emissions by only 7%. However, nonparticipants increased their pace of reductions from an average of 3.5% per year between

TABLE 2-4. Chemical-Specific Reduction in Releases

Names of 33/50 chemicals	1988 (million pounds)	1990 (million pounds)	1995 (million pounds)	% reduction relative to 1988	% reduction relative to 1990
Benzene	36.8	30.1	11.7	68	61
Carbon tetrachloride	5.3	2.9	1.2	77	59
Chloroform	29.8	26.4	12.7	57	52
Dichloromethane	155.4	112.9	71.0	54	37
Methyl ethyl ketone	171.8	156.9	77.0	55	51
Methyl isobutyl ketone	45.0	34.4	24.0	47	30
Tetrachloroethylene	42.5	27.9	12.0	72	57
Toluene	367.5	294.5	169.0	54	43
1,1,1-Trichloroethane (TCA)	200.9	182.4	24.5	88	87
Trichloroethylene	62.6	44.0	27.2	57	38
Xylenes	212.9	177.5	119.4	44	33
Cadmium and cadmium compounds	1.8	1.8	2.4	-31	-30
Chromium and chromium compounds	71.2	63.2	51.5	28	19
Cyanide compounds	12.0	8.5	9.7	20	-14
Lead and lead compounds	60.9	78.6	44.7	27	43
Mercury and mercury compounds	0.32	0.2	0.2	25	0
Nickel and nickel compounds	19.6	22.8	14.3	27	37
Total	1496.3	1265.1	672.5	55	47

1988–1990 to 7% per year between 1990 and 1995. Participants, on the other hand, maintained their pace of reductions at an average of about 10% per year between 1988–1990 and between 1990–1995.

The two ozone-depleting substances among the 33/50 chemicals had the largest percentage reductions from 1990 to 1996. While releases of other 33/50 chemicals decreased from 40% from 1990 to 1995 and 9% in 1996, those of carbon tetrachloride and of TCA decreased 86% during 1990–1995 and 54% in 1996. This suggests that policies other than the 33/50 program played an important role in reduction of the 33/50 releases. Reduction in other 33/50 releases amounted to 40% of the levels in 1990 and 50% of the levels in 1988. If these two ozone-depleting chemicals are excluded, reduction in emissions of the remaining chemicals was still 50% as compared to 1988 levels but only 40% as compared to 1990 levels. The largest absolute reduction among these chemicals was reported for toluene, which had the greatest releases among the 33/50 chemicals in 1988. It continued to have the greatest releases among these chemicals in 1995. The releases of cadmium and cyanide compounds increased during the program years (Table 2-4).

Comparing the rate of decline of 33/50 releases relative to those of other TRI chemicals over the period 1988–1995, one finds that while the former declined by 55%, the latter declined by 36%. The corresponding rates of decline between 1991 and 1995 were 47% and 25%, respectively, suggesting that the program did contribute to a faster decline of program chemicals as compared to other TRI chemicals (U.S. EPA 1999).

Pollution Prevention

While the 33/50 program emphasized reducing toxic chemicals through source reduction, it did not require participants to use only source reduction methods to reduce 33/50 releases. Since 1991, the TRI has required reporting on chemicals recycled on and off-site; those combusted for energy recovery on and off-site; those treated on and off-site; and those released on-site and sent off-site for disposal. Together, these data are referred to as production-related waste. Trends in the overall level of 33/50 chemical waste generated and in the relative importance of alternative waste management techniques employed can be seen in Table 2-5.

On-site and off-site releases (during disposal) targeted specifically by the 33/50 program declined by about 34% between 1991 and 1995. Nevertheless, total production-related waste of 33/50 chemicals remained essentially unchanged during the program period, declining by only 2% between 1991 and 1995. This is because other categories of waste increased from 4.7 billion to 5.5 billion pounds. More specifically, off-site recycling increased by 17%, while on-site treatment increased by 54%. These increases were partly offset by a 2% reduction in on-site recycling and by a 13% reduction in on-site energy recovery. This suggests that facilities achieved their reductions in on-site and off-site releases primarily by increasing on-site treatment and off-site recycling (U.S. EPA 1999). Although the production-related waste of 33/50 chemicals declined only slightly, that of other TRI chemicals increased by 2% over the same period, suggesting that production-related waste of the 33/50 chemicals may have been higher in the absence of the program.

Since 1991, each facility also was required to report any of eight different activities they adopted to reduce pollution at source for each toxic chemical reported by them to TRI in a given year.[9] The number of 33/50 chemicals released by each facility summed over all facilities reporting to the TRI is the number of opportunities for adopting source-reduction activities. Of the total number of opportunities for reducing 33/50 releases in 1991, 33% reported adopting a source reduction activity; the corresponding estimate for 1995 was 29%. In contrast, only 22% of the opportunities for reducing releases of other TRI chemicals were utilized in 1991; the corresponding estimate for 1995 was 20%. Adoption rates were higher, therefore, for 33/50 releases than for other TRI chemicals. The three TRI chemicals for which the largest number of facil-

TABLE 2-5. Total 33/50 Chemical Waste Generated

	1991 (billion pounds)	*1995 (billion pounds)*	*% change from 1991– 1995*
On-site recycling for 33/50 chemicals	2.56	2.52	-1.6
Off-site recycling for 33/50 chemicals	0.64	0.75	17.2
On-site energy recovery for 33/50 chemicals	0.71	0.62	-12.7
Off-site energy recovery for 33/50 chemicals	0.24	0.84	2.5
On-site treatment for 33/50 chemicals	0.48	0.74	54.2
Off-site treatment for 33/50 chemicals	0.08	0.079	1.4
On-site and off-site releases of 33/50 chemicals	0.88	0.58	-34.1
Total production-related waste of 33/50 chemicals	5.6	5.5	-1.8
Number of opportunities for source-reduction activities for 33/50 chemicals	26,654	21,140	
Adoption rate for source-reduction activities (% of opportunities)	33%	29.3%	

Source: U.S. EPA (1999).

ities (in absolute and percentage terms) reported source reduction were three 33/50 chemicals: TCA, toluene, and xylenes. The most commonly adopted source-reduction activity for reducing 33/50 releases, as well as for other TRI chemicals, was good operating practices followed by process modifications. The facilities adopting source-reduction activities were those generating relatively larger volumes of production-related waste of those chemicals. In 1991, these facilities accounted for 45% of total production-related waste of 33/50 chemicals and 33% of production-related waste of other TRI chemicals.

Thus evidence on the extent to which the 33/50 program promoted pollution prevention is mixed. While firms did adopt more source-reduction activities for 33/50 chemicals, EPA (1999) did not analyze whether program participants did more than nonparticipants. Moreover, the increase in recycling and treatment suggests that releases prevented through those source reduction activities may have been limited.

Selected Case Studies of Participants

Case studies show that some participants did make significant efforts to reduce their 33/50 releases. Baxter and Lilly were two multinational firms that spent $80 million and $10 million, respectively, to reduce TRI releases, especially 33/50 releases. Lilly used two of the 33/50 chemicals, toluene and methylene chloride, in large quantities; the latter accounted for more than 70% of Lilly's total releases. Lilly pursued three strategies for reducing these releases: new manufacturing technologies and practices to reduce evaporation and increase containment within the facility; central collection and incineration; and reduction in the use of methylene chloride in its new products. The last involved identifying non-toxic substitutes and less volatile solvents for methylene chloride and other toxic chemicals. As a result, Lilly reduced its 33/50 releases by 81% between 1988 and 1995 (Prakash 2000).

Baxter used eight of the 33/50 chemicals; in particular, methyl ethyl ketone and TCA in significant quantities. Adverse publicity following the release of the TRI led Baxter to take aggressive steps to reduce its toxic releases. By 1991, Baxter had ongoing programs to reduce its 33/50 releases and committed to an even higher standard than was required by the 33/50 program. It committed to an 80% reduction goal, not only for 33/50 releases but for CFCs and for hazardous air pollutants by 1995 for all of its facilities. In reducing these releases, Baxter followed EPA's hierarchy of pollution control methods: awarding the highest priority to source reduction methods, followed by recycling and treatment. Baxter reduced its releases of TCA by more than 80% by 1990 by modifying process leach tanks and instituting operational changes. It then developed a new technology and switched to an aqueous-based cleaning process that eliminated TCA emissions by 1991. Emissions of methyl ethyl ketone were reduced by 60% between 1998 and 1994 by modifying process and reducing fugitive losses. Total 33/50 releases fell by 96% from 0.9 million pounds in 1988 to 0.037 million pounds in 1995. During this period, companywide production increased by 20% (Prakash 2000).

3M was the largest emitter of 33/50 releases in 1988 and committed to reduce these releases by 60% by 1995. It later modified this goal to 70% to be met by the end of 1992. Reduction in 33/50 releases was a part of the long-established pollution prevention programs at 3M, but 3M differed from EPA in its definition of pollution prevention. It considered environmentally sound reuse and recycling to be a method for preventing pollution. In 1989, 3M's releases of methyl ethyl ketone represented 12% of national releases. 3M developed techniques to capture both methyl ethyl ketone and toluene and recycle them for reuse. By 1991, it reduced its 33/50 releases by 54% (Hopkins and Allen 1994/1995).

Dow Chemical Company is one of the 10 largest chemical producers in the world. In 1991, it pledged to reduce its toxic air emissions by 71% at six plants by 1994 and focused on specific 33/50 chemicals. The company found the 33/50 program to be consistent with its own goals under its Waste Reduction Always Pays program, which focused on reducing wastes through elimination, reclamation, treatment and destruction, and secure landfill disposal. Like 3M, it viewed pollution prevention as encompassing recycling. In 1988, it initiated a recycling project to reduce emissions of methylene-chloride in wastewater. It installed equipment that improved separation of the solvent from wastewater, allowing a better recovery rate for reuse. In 1988, Dow released 1 million pounds of this chemical. This was reduced by 41% by 1991. Its total 33/50 releases decreased from 6 million pounds to 2.4 million pounds between 1988 and 1991 (Hopkins and Allen 1994/1995).

These case studies describe the willingness of companies to spend considerable resources to eliminate their 33/50 releases and possibly save some resources by allowing reuse of chemicals from the wastestream (Hopkins and Allen 1994/1995). They also show that companies were motivated to participate by a desire to be environmental leaders, to demonstrate that cooperative approaches between industry and government could work, to avoid more stringent command-and-control regulations in the future, and to reduce the adverse publicity associated with being listed in the TRI. In addition to reducing releases, the 33/50 program had other benefits, such as leading to improved communication about pollution within participating firms. It led senior executives in companies to communicate more with employees closest to the problem. The program led to heightened awareness about the costs of waste generation and disposal that could be avoided and provided an incentive for firms to make efforts to reduce their emissions by setting their own goals and achieving them using methods chosen by them.[10]

Motivations for Participation in the 33/50 Program

Several academic studies have examined the factors that increased the likelihood of a firm participating in the 33/50 program.[11] These studies show the importance of public recognition provided by the program as a motivator for participation. Firms that are primarily producing final goods and in closer contact with consumers or in industries with higher advertising expenditures per unit of sales and, therefore, more visible to consumers, were found to be more likely to participate in the 33/50 program (Arora and Cason 1995, 1996; Celdren et al. 1996; Khanna and Damon 1999). Videras and Alberini (2000), Sam and Innes (2005), and Gamper-Rabindran (2006), on the other hand, found that proximity to final consumers had an insignificant effect on program participation. These studies found that participation in the program was motivated by a desire to offset adverse publicity from previous boycotts (Sam and Innes 2005); a desire for a positive public image by firms that published annual environmental reports (Videras and Alberini 2000); and a desire by publicly owned firms to appeal to investors likely to prefer firms facing fewer environmental liabilities and lower costs of compliance in the future (Gamper-Rabindran 2006). Concerns about adverse publicity could also explain the finding that firms producing larger volumes of 33/50 releases or other toxic releases were more likely to participate in the 33/50 program (Arora and Cason 1996; Khanna and Damon 1999; Sam and Innes 2005; Gamper-Rabindran 2006).

Additionally, the threat of liabilities, more stringent enforcement of existing regulations, and high costs of compliance with anticipated regulations also motivated firms to participate in the 33/50 program. While Khanna and Damon (1999) and Videras and Alberini (2000) found that firms that were listed as potentially responsible parties for a relatively large number of Superfund sites were more likely to participate in the 33/50 program, Sam and Innes (2005) that this was the case only for large firms. They also found that firms with higher rates of inspections in the past were more likely to have participated in the program, while Gamper-Rabindran (2006) found this to be the case for firms in the chemical industry only. Videras and Alberini (2000) found that firms subject to corrective actions for violations of the Resource Conservation and Recovery Act were more likely to participate in the 33/50 program. Additionally, Khanna and Damon (1999) and Gamper-Rabindran (2006) found that firms in the chemical industry that had a high share of hazardous air pollutants in total toxic releases, and, therefore, faced a stronger threat of high costs of compliance with anticipated regulations, were more likely to participate in the 33/50 program.

Contrary to expectations based on the raw data that parent companies signing up first for the 33/50 program would be those that had made greater progress in reducing their 33/50 releases prior to the start of the program in 1992, all of the above cited studies show that the extent of reduction achieved by firms prior to 1991 did not have a statistically significant effect on the program participation decision after controlling for the effects of the other factors that differed across firms. This was also observed to be the case by INFORM (1995) using a matched-pair analysis method.[12] The study found that facilities that participated had achieved smaller reductions by 1991 as compared to matching facilities that did not participate.

High costs of participation, as proxied by the ratio of 33/50 releases to all TRI releases, is found to have a statistically significant negative impact on incentives for participation in the chemical industry (Khanna and Damon 1999; Gamper-Rabindran 2006) and in the fabricated metal industry (Gamper-Rabindran 2006). This suggests that firms relying more heavily on 33/50 chemicals may have found it more costly or technically difficult to find alternative substitutes for these chemicals and were therefore less likely to participate in a program aiming to reduce the use of those chemicals. A few studies also examined the effect of the financial health of the firm and found it not to be a strong deterrent to participation in the 33/50 program (Celdran et al. 1996; Arora and Cason 1995). The survey of firms by Celdran et al. (1996) suggests that for some firms it was technically infeasible to modify their processes and therefore to participate in the program.

In summary, these findings suggest that a desire for positive publicity and to create an environmentally friendly public image, for deterring boycotts/offsetting the adverse effects of previous boycotts, and for reducing the threat of enforcement and liabilities were important motives for 33/50 program participation.

Analysis of Program Achievements

External analysts of the 33/50 program have had divergent views about its performance. Protagonists such as Richard Zanetti (1995), editor of *Chemical Engineering*, reflecting the views of the chemical industry, wrote that "33/50 has been one of the smartest things EPA had done regarding the environment—the government's role in 33/50 is small as it should be. In fact, EPA manages the entire program with just a handful of staffers.[13] I would guess that taxpayer return on investment for 33/50 is light years ahead of that from traditionally run environmental programs."

Others, such as the GAO (1994) and various pro-environment groups, noted that not all of the reported reductions could be attributed to the program. From the data reported in U.S. EPA (1999) and presented in Table 2-3, it can be observed that about 28% of the total reduction (of 824 million pounds by 1995) occurred between 1988 and 1990, prior to the initiation of the program. Additionally, firms that did not participate in the program reported 22% of that reduction (equivalent to 187 million pounds). Therefore, only 50% of the reduction reported as having been achieved by the program was estimated to be due to the participants' efforts between 1991 and 1995. In support of this, the Citizen Fund (1990) found that 83% of all facilities had started to make reduction in 33/50 releases even prior to 1991, and INFORM (1995) found that 31% of program participants had initiated reduction activities prior to 1991.

The GAO (1994) recommended that EPA only consider reductions achieved after 1991 as a contribution of the program. It concluded that without counting the reductions of the nonparticipants and the achievements prior to 1991, the program would not have met its interim goal. Moreover, the reductions achieved not only in 33/50 releases but also in other TRI releases suggest that other secular changes (such as changes in production levels) besides the program were influencing these reductions. It also is possible that the choice of 1988 as the baseline level stimulated greater participation and, thus, greater reduction in releases than might have been forthcoming with a 1991 baseline.

One of the key objectives of the program was to encourage firms to prevent pollution at source to achieve the targeted emissions reduction. Since EPA did not require companies to report the extent to which reductions in 33/50 releases were achieved due to source reduction as opposed to treatment or other methods, the GAO (1994) concluded that the program's impact on pollution prevention was uncertain. Like the GAO, many environmentalists also felt that efforts to cut 33/50 releases were underway prior to 1991 when the program was initiated (BNA 1994). Additionally, they were concerned about the accuracy of the data reported by firms about their releases to the TRI, since EPA was not monitoring participating firms. They also disagreed with EPA's definition of pollution prevention. While EPA defined it as reduction of releases at the point of departure, environmental groups would have preferred to define it as reduction in chemicals used by a facility and would have liked EPA to monitor chemicals from the point they reach the facility.[14]

Several academic studies have sought to isolate the impact of the 33/50 program on 33/50 releases using a with-and-without comparison of environmental performance by controlling for other factors that also could have influenced releases.[15] Program impact cannot be determined simply by comparing the difference in percentage reductions in releases achieved by participants and nonparticipants. Such a comparison would be appropriate when participants and nonparticipants are identical in all respects other than in their participation in the program. Additionally, since this is a voluntary program and firms self-select as participants, it is necessary to control for the possibility that firms that participated in the program had an inherent propensity to reduce emissions more than nonparticipants; for example, due to a management with green preferences. The studies discussed below incorporate these considerations and control for the possibility that the decision of firms to participate in the program and the impact of participation on their releases is likely to be jointly determined and influenced by many of the same observable and unobservable factors.

The earliest study by INFORM (1995) found that the percent change in 33/50 releases between 1988 and 1992 was not much higher for facilities of participating companies as compared to those of non-participating companies. The study did not evaluate the impact of the program on 33/50 releases by participants after the program started in 1991. It did, however, find that participating facilities showed a larger rate of reduction in releases as compared to nonparticipants after 1991.

Khanna and Damon (1999) evaluated the impact of the 33/50 program on firms in the chemical industry using data for the period 1988–1993. They found that large and increasing potential liabilities under the Superfund Act and costs of compliance with the proposed National Emissions Standards for Hazardous Air Pollutants (NESHAP) had statistically significant deterrent effects on the level of 33/50 releases generated by a firm. The level of 33/50 releases was positively affected by the volume of sales and negatively by the sales per unit assets ratio, a measure of the idle capacity in a firm. Firms with newer assets or those with a high rate of asset replacement have lower 33/50 releases. However, even after controlling for the direct effects of these variables on the level of releases, they found that program participation led to a statistically significant decline in 33/50 releases. A 10% increase in the probability of participation of a firm reduced its 33/50 releases by 5.1%. They found that the program led to an expected reduction of 28% in 33/50 releases over the period 1991–1993 relative to the pre-program level of releases in 1990. The study also found that program participation had a significant negative impact on emissions to all major media to which they were discharged: air, land, water, and disposal facilities. This suggests that the design of the program, which was not media-specific, encouraged integrated environmental management and reduction in total emissions and did not create incentives for cross-media substitution. The study found that program participation had a negative impact on releases of other TRI chemicals, but it was less significant than on 33/50 releases, suggesting that program participation had scope effects that led to a reduction in releases of other chemicals as well.

Maxwell, Lyon, and Hackett (2000) identified political-economic factors that explain reductions in 33/50 releases per unit of shipments. Using state-level data for the period 1988–1992, they found that states with high environmental group membership experienced larger reductions in releases. They attribute this effect to relatively low marginal abatement costs for firms, a high value for abatement by consumers, and low costs of organizing for consumer interest groups. In such states, the threat of mandatory regulation is likely to be high, while the cost of self-regulation for firms is likely to be low, and hence firms are more likely to engage in voluntary emissions reductions. Furthermore, they hypothesize that firms have an incentive to collusively choose their voluntary abatement levels to preempt regulation. As the number of firms increases, the potential for free-riding increases and the extent of voluntary abatement is likely to decrease. They find empirical support for this hypothesis, since states with a smaller number of plants that produce 33/50 releases experienced greater levels of reduction in these releases, possibly because firms in these states found it easier to overcome free-rider problems and coordinate abatement by plants in an industry. Lastly, states with larger manufacturing shipments reduced releases faster, possibly because industry in these states was more concerned about stringent regulations in the future and also more able to afford larger abatement levels. These results show that other factors besides program participation were important in explaining the reductions in 33/50 releases by all firms in general.

More recently, Sam and Innes (2005) examined the effect of the 33/50 program on releases and on the frequency of inspections of firms that were invited to participate in the 33/50 program in 1991 from seven different two-digit Standard Industrial Classification (SIC) codes. Using data for 1989–1995, they examined the impact of the program on 33/50 releases from 1992 onwards.[16] They found that program participation did lower 33/50 releases starting in 1992 and that this effect persisted until 1995. Additionally, they found that firms that were in more concentrated industries and had higher R&D expenditures were more likely to have lower 33/50 releases, possibly to set high environmental standards and gain a competitive advantage by raising the costs of abatement for rival firms. They find evidence that firms may have reduced 33/50 releases to preempt a demand for more stringent regulations by environmental interest groups and to deter boycotts in states with a large presence of environmental interest groups. They also analyzed the effects of program participation on the frequency with which firms are inspected and found that it led to a significant decline (by 17% relative to sample average) in inspection rates from 1993 to 1995.

Gamper-Rabindran (2006) analyzed the change in releases due to the 33/50 program over the period 1990–1996 using data for all facilities eligible to participate in the 33/50 program and that are covered by the Clean Air Act reporting requirements. Only 15 of the 17 program chemicals were included in the analysis, excluding the two ozone-depleting chemicals whose phase-out was mandated by the Clean Air Act. Industry-specific analysis showed that the impact of the program varied across sectors; the program led to a statistically significant reduction in releases in the paper and fabricated metals sector only, while it led to a significant increase in releases in the chemicals and primary metals sectors (1991–1995). The effects of the program on releases from the electric; transport; rubber and stone sectors were statistically insignificant. Releases in the fabricated metals sector fell by 50% of participant's pre-program emissions over the period 1991–1995, while in the paper industry, the corresponding decrease was by 95% over the period 1991–1996. Program participants from the chemical industry increased releases by 70%, while those in the primary metals industry increased releases by 95% (1991–1995). The effects of the program on toxicity-weighted releases were found to be similar. Additionally, the program was found to have led to an increase in toxicity-weighted off-site transfers to recycling in all industries except the primary metals, chemical, and transport industries over the 1991–1995 period. The paper also found that program participants reduced their toxicity-weighted 33/50 releases in politically active counties with high voter participation rates.[17]

Because the 33/50 program was voluntary, it is reasonable to believe that profit-maximizing firms would have participated in it only if they expected to achieve higher profits than otherwise. Participation may lead to higher profits for firms by focusing their attention toward increasing efficiency in the use of 33/50 chemicals, as well as by reducing future environmental liabilities and future costs of compliance with mandatory regulations. A reduction in the pollution generated by firms also could increase consumer goodwill, improve the public image of a firm, and increase investor confidence. Investors may perceive a firm that commits to reducing its toxic pollution ahead of time by flexible methods as gaining a strategic advantage over its competitors. While many of the expenditures on pollution control, particularly on capital equipment, are likely to be incurred in the short term, most of the benefits only may be realized in the long run. The impact of program participation on current and expected long-run profitability of firms, therefore, may differ.

To examine this, Khanna and Damon (1999) analyzed the impact of the program on the financial performance of firms, measured by the return on

TABLE 2-6. Impact of the 33/50 Program on Releases

Study	Period of study; dependent variable	Sample size	Impact of program participation on releases	Other effects of program participation
Khanna and Damon (1999)	1988–1993; level of 33/50 releases	123 firms in SIC 28	Statistically significant negative impact (-) 28% (1991–1993) relative to 1990 levels	Program participation reduced return on investment by 1% and increased excess value/sales by 2%.
Sam and Innes (2005)	1989–1995; level of 33/50 releases	319 firms from SICs 20–39 invited to participate in the program in 1991 with 3 or more years of complete data (1988–1995)	Statistically significant negative impact on 33/50 releases from participation between 1992–1995 period.	Statistically significant negative impact on inspections under the Clean Air Act (1993–1995).
Gamper-Rabindran (2006)	1990–1996; change in 33/50 releases (non–ozone depleting only); 1991–1995	Facility level SIC 26: paper: 216 SIC 34: fabricated metal: 1,043 SIC 33: primary metal: 488 SIC 28: chemical: 791 SIC 36: electronics: 358 SIC 37: transport: 37	Statistically significant negative (-) 51% and (-) 95% for SIC 34 and 26. Statistically significant positive (+) 170% for SIC 28; (+) 97% for SIC 33; insignificant effect on other sectors.	Heath-indexed emissions: (-89%) and (-390%) for SICs 26 and 34 (+180%) and (+110%) for SICs 28 and 33; insignificant effect on others. Statistically significant positive effect on off-site recycling (except SIC 28).

Note: SIC = Standard Industrial Classification.

investment (ROI) and the excess value per unit sales (EV/S), measured by the market value of a firm's book value of assets/sales. The ROI is an accounting measure of financial performance and a measure of the success of past and current investments, while EV/S is based on market value data and provides a future-oriented measure of a firm's economic performance reflecting current expectations about a firm's ability to achieve profits in the future. Their analysis shows that the program led to a statistically significant expected decline of 1.2% in average ROI of firms but an expected increase of 2.2% in EV/S over the period 1991–1993. This suggests that while the immediate impact of program participation on firms was negative, possibly due to the increased expenditures on pollution control, investors expected such firms to be more profitable in the long run.

Conclusion

The 33/50 program represented an innovative experiment in which firms cooperated with regulators by voluntarily reducing pollution and going beyond compliance. Chris Tirpak of EPA's 33/50 program staff remarked, "Companies cut more than 750 million pounds of pollution in the time it would take EPA to write a regulation. . . . We were stunned that companies signed up to make these reductions publicly . . . [and took the] risk in admitting [they were] a big polluter. But companies even signed up for 90 percent reductions and some companies made it. All we did was set the national goals and identify the targets. The companies chose their own methods."[18] By the time the 33/50 program ended in September 1996, it had spawned 80 more voluntary programs initiated by EPA, making it the "harbinger of a changing regulatory rationale" (Zanetti 1995). Firms welcomed the program because it allowed them to preempt more command-and-control-type regulations. In 1996, 3M nominated the 33/50 EPA team for the Vice President's Hammer Award, commending it for exchanging "blue ribbons for red tape." This was the first time a firm had nominated a government regulatory agency for that award.

It is important to recognize that the 33/50 program implicitly represented a carrot-and-stick approach toward environmental protection. It provided incentives for voluntary reduction, such as public recognition to firms and flexibility in methods and extent of pollution control, while having a credible threat of mandatory regulations if voluntary efforts were not successful. The incentives provided by the program were "soft" incentives and neither monetary incentives nor explicit regulatory relief were promised in exchange for voluntary action. In the absence of voluntary action, however, regulations were believed to be inevitable; some of the 33/50 chemicals were already facing impending regulations under the Clean Air Act Amendments of 1990. Moreover, the program was accompanied by publicly available toxic release information that allowed firms, regulators, and the public to keep track of progress made by firms toward achieving the goals of the program. This provided an implicit threat of adverse publicity for firms that had large 33/50 releases and were not making proactive efforts to reduce them. The 33/50 program is unique among the numerous other voluntary programs that were established in the United States after it. It has been the only voluntary program with clearly defined numerical goals, an established baseline, and a mandatory requirement for reporting releases to track performance of the program.

The program was successful in inducing participation by firms that accounted for more than 60% of the 33/50 releases in 1988. Empirical studies show that the desire to be publically recognized, to avoid adverse publicity, and to reduce the threat of liabilities, frequency of regulatory enforcement, and cost of future regulations motivated participation in the 33/50 program. Moreover, 33/50 releases did decline considerably between 1988 and 1995. Firm-level analysis of the effects of the program on the releases of all 17 chemicals suggests that the program did have a negative impact on releases. Facility-level and sector-level analysis focusing only on the 15 ozone-depleting chemicals indicates, however, that the effects of the program varied across sectors. For some sectors, program participation was found to have increased or had an insignificant effect on releases. Moreover, the extent to which the program motivated source-reduction of pollution is not clear. The raw data suggest that while firms reduced their on-site releases and off-site transfers of 33/50 chemicals, they increased recycling and treatment of these chemicals rather than reducing the generation of these chemicals at source. By specifying goals of the program in terms of reduction in the 33/50 releases and transfers and linking public recognition incentives to those reductions, the program created incentives for firms to find the cheapest way to achieve those reductions. An inability to measure the extent of pollution reduced at source may have limited incentives of firms to undertake it.

Overall, the experience with the 33/50 program shows that firms respond to incentives that allow them to differentiate themselves from rivals and signal their environmental commitment to the public and to regulators, thereby improving their reputation and gaining goodwill. It also demonstrates the importance of having a strong regulatory framework and an environmentally aware public that provides a credible threat of high-cost mandatory regulations and adverse publicity if voluntary actions are not forthcoming. Finally, it highlights the need for clearly defined goals with measurable indicators of progress toward achieving those goals.

References

Arora, S., and T. N. Cason. 1995. An Experiment in Voluntary Environmental Regulation: Participation in EPA's 33/50 Program. *Journal of Environmental Economics and Management* 28: 271–286.

———. 1996. Why Do Firms Volunteer to Exceed Environmental Regulations? Understanding Participation in EPA's 33/50 Program. *Land Economics* 72: 413–432.

BNA (Bureau of National Affairs). 1994. EPA 33/50 Pollution Prevention Program Labeled "Sham" by Environmental Groups. *BNA National Environment Daily*, June 3.

Celdran, A., H. Clark, J. Hecht, E. Kanamaru, P. Orantes, and M. Santaello Garguno. 1996. The Participation Decision in a Voluntary Pollution Prevention Program: The U.S. EPA 33/50 Program as a Case Study. In *Developing the Next Generation of the U.S. EPA's 33/50 Program: A Pollution Prevention Research Project*, edited by H. Clark. Durham, NC: Nicholas School of the Environment, Duke University.

Citizen Fund. 1994. Pollution Prevention or Public Relations? Washington, DC: Citizen Fund.

Gamper-Rabindran, Sharti. 2006. Did the EPA's Voluntary Industrial Toxics Program Reduce Emissions? A GIS Analysis of Distributional Impacts and a By-Media Analysis of Substitution. *Journal of Environmental Economics and Management* 52(1): 391–418.

GAO (United States General Accounting Office). 1994. *Toxic Substances: EPA Needs More Reliable Source Reduction Data and Progress Measures*. GAO/RCED-94-93.

Hart, S. 1995. A Natural Resource-Based View of the Firm. *Academy of Management Review* 20: 985–1014.

Hopkins, L., and D.T. Allen. 1994/1995. Voluntary P2 Initiatives: Industry Responds to EPA's 33/50 Program. *Pollution Prevention Review* Winter.

INFORM. 1995. *Toxics Watch 1995*. New York: INFORM.

Khanna, M., and L. Damon. 1999. EPA's Voluntary 33/50 Program: Impact on Toxic Releases and Economic Performance of Firms. *Journal of Environmental Economics and Management* 37(1): 1–25.

Kirschner, E.M. 1995. Industry Sees Maturation, Contradiction in EPA's Quarter-Century History. *Chemical and Engineering News*, October 30.

Maxwell, J.W., T. P. Lyon, and S.C. Hackett. 2000. Self-regulation and Social Welfare: The Political Economy of Corporate Environmentalism. *Journal of Law & Economics* XLIII (2): 583–618.

New York Times. 1991. The Nation's Polluters—Who Emits, What and Where. October 13, p. F10.

Popoff, F.P. 1992. Going Beyond Pollution Prevention. The Conference Board, *Championing the Global Environment*, p. 17.

Porter, M.E., and C. van der Linde. 1995. Toward a New Conception of the Environment-Competitiveness Relationship. *Journal of Economic Perspectives* 9: 97–118.

Prakash, A. 2000. *Greening the Firm: The Politics of Corporate Environmentalism*. Cambridge, UK: Cambridge University Press.

Reinhardt, F. 1999. Market Failure and the Environmental Policies of Firms. *Journal of Industrial Ecology* 3(1).

Sam, A.G., and R. Innes. 2005. Voluntary Pollution Reductions and the Enforcement of Environmental Law: An Empirical Study of the 33/50 Program, manuscript, Department of Agricultural and Resource Economics, University of Arizona, Tucson.

U.S. EPA (Environmental Protection Agency). 1990. *Reducing Risk: Setting Priorities and Strategies for Environmental Protection.* Washington, DC: U.S. EPA.

———. 1992. EPA's 33/50 Program Second Progress Update. Office of Pollution Prevention and Toxics (TS-792A). Washington, DC: U.S. EPA.

———. 1994. EPA's 33/50 Program Fifth Progress Update. Office of Pollution Prevention and Toxics (7408). Washington, DC: U.S. EPA.

———. 1999. 33/50 Program: The Final Record. EPA-745-R-99-004, Office of Pollution Prevention and Toxics (7408). Washington, DC: U.S. EPA.

———. 2000. Taking Toxics Out of the Air. EPA-452/K-00-002. Research Triangle Park, NC: U.S. EPA, Office of Air Quality, Planning and Standards.

Videras, J., and A. Alberini. 2000. The Appeal of Voluntary Environmental Programs: Which Firms Participate and Why? *Contemporary Economic Policy* 18(4): 449–461.

Zanetti, R. 1995. A Voluntary Cleanup Program that Works. *Chemical Engineering,* August, p. 5.

Notes

1. http://www.fda.gov/cdrh/leveraging/2.html (accessed December 14, 2006).

2. http://www.epa.gov/history/topics/programs/01.htm (accessed November 27, 2006).

3. William K. Reilly, "Aiming Before We Shoot: The Quiet Revolution in Environmental Policy," address to the National Press Club on Sept. 26, 1990. http://www.epa.gov/history/topics/risk/02.htm (accessed November 27, 2006).

4. http://www.epa.gov/history/topics/risk/02.htm (accessed November 27, 2006)..

5. Facilities could obtain a six-year extension to comply with the MACT standards if they participated in an Early Reductions Program established by the EPA. To do this they had to significantly reduce their emissions before the EPA proposed the standards.

6. http://www.fda.gov/cdrh/leveraging/2.html.

7. The EPA had proposed requiring firms to provide information to the TRI on their source reduction and recycling activities, the quantities of waste prevented by these activities, uses of specific chemicals, amounts of on-site and off-site transfers, and methods of on-site waste treatment. By collecting information on the quantities of chemicals that would have entered the wastestream if companies had not implemented source reduction activities, the EPA could track the extent to which firms were reducing pollution at source as opposed to using other methods to manage their releases. However, the Office of Management and Budget pointed out that the Pollution Prevention Act required the reporting only on the *number of practices* used to achieve source reduction and not on the *quantities of waste* resulting from those practices; hence the lack of information on the latter (GAO 1994).

8. This estimate of other TRI chemicals does not include delisted chemicals, chemicals added in 1990, 1991, 1994, and 1995, ammonia, hydrochloric acid, and sulfuric acid. It also does not include the 33/50 releases.

9. These activities consist of: 1) changes in operating practices; 2) materials and inventory control; 3) spill and leak prevention; 4) raw material modifications; 5) equipment and process modifications; 6) rinsing and draining equipment design and maintenance; 7) cleaning and finishing practices; and 8) product modifications.

10. http://www.fda.gov/cdrh/leveraging/2.html.

11. These include: Arora and Cason (1995, 1996); Celdren et al. (1996); Khanna and Damon (1999); Alberini and Videras (2000); Sam and Innes (2005); and Gamper-Rabindran. Among these, the analysis by Khanna and Damon (1999) is based on firm-level data for the chemical industry only, while the analysis by Gamper-Rabindran is based on facility-level data disaggregated across six different industries: paper, fabricated metal, primary metal, chemical, electronic and electric, and transport. The other studies, with the exception of Arora and Cason (1996), which used facility-level data, use firm-level data pooled from many different industries.

12. The study used data for the period 1988–1992 and a matched-pair analysis technique to match a participating facility with a nonparticipating facility. Firms were matched depending on whether they produced the same chemical, performed a similar activity, were located in the same Standard Industrial Classification (SIC) code and state, and had total releases and transfers of TRI chemicals that were within plus or minus 50% of each other.

13. The 33/50 program was implemented with a relatively modest budget of $400,000 and a staff of six at headquarters and two regional staff.

14. http://www.fda.gov/cdrh/leveraging/2.html.

15. Khanna and Damon (1999); Sam and Innes (2005); Gamper-Rabindran.

16. Instead of defining participation as a 0/1 dummy variable (equal to one for a participant and 0 otherwise) for 1991 only and examining the effect of participation in 1991 on releases thereafter, they determine the incremental effect of each year's participation on 33/50 releases by constructing four participation variables. Each of these is defined as a 0/1 variable for a firm for each of the remaining years of the program.

17. The paper does not provide details of the direct effects of other explanatory variables capturing consumer, investor, regulatory, and community pressure on the change in 33/50 releases.

18. http://govinfo.library.unt.edu/npr/initiati/common/reinvreg.html (accessed December 14, 2006).

3

Japan's Keidanren Voluntary Action Plan on the Environment

Masayo Wakabayashi and Taishi Sugiyama

In this chapter, we provide an overview of Japan's 1997 Keidanren Voluntary Action Plan on the Environment (Nippon Keidanren 1997). This program was initiated by the Keidanren, the most prominent business association in Japan. There are two key features of the Keidanren plan. First, it is wide in its coverage. Keidanren encompasses large enterprises drawn mostly from the industrial, energy (electricity, petroleum, and gas), construction, commercial, and transport sectors. This broad coverage reflects the tradition of horizontal egalitarianism among industries, which is also a prominent feature of Japanese business culture. Second, the government is deeply involved in the implementation process of the Keidanren plan. Although it is a voluntary agreement, the plan has several features, such as close linkage with government planning, that compel industries to comply.

Certain characteristics of the Keidanren plan, such as the emphasis on voluntary environmental agreements, are less common in the climate policies of other countries. In order to share this concept of voluntary agreement with readers, we begin with an overview of the traditional development of environmental policy in Japan, focusing on the relationship between industry and government. Next, we review the Keidanren plan itself and highlight its distinctive features. Finally, we identify several of the existing compliance mechanisms at work in the Keidanren plan in the absence of legal enforcement.

Environmental Policy in Japan

Problems associated with pollution have been documented in Japan since the nineteenth century, but economic progress and development had priority over pollution control and the protection of environmental quality. However, a series of incidents of pollution in the 1950s and 1960s led to increased public pressure on industries to act in a socially responsible manner. This spawned important legislative and regulatory measures, such as decontamination efforts, the establishment of environmental standards, and requirements for compensation for the damages caused by environmental pollution. However, this first wave of national pollution-control laws required very little of industry in contrast to international standards at the time. As a result, local citizens and regional governments placed still further pressure upon industry to strengthen pollution-control policies. A series of negotiations between local stakeholders and regional governments followed, resulting in almost every major enterprise across all sectors conceding to the demands placed upon them. These compliance arrangements, termed "Pollution Prevention Agreements," typically involved the adoption of best-available technologies (Investigation Committee on Japan's Air Pollution Experience 1997).

As a result of these arrangements, the level of pollution-control performance in Japan has improved dramatically. In the 1970s, there was massive investment in pollution-control technologies. For example, firms in the power sector installed scrubbers to clean up harmful emissions of sulfur oxides. In comparison, it took more than a decade in Europe (under the Convention on Long-Range Transboundary Air Pollution) and two decades in the United States (under the Clean Air Act and Clean Air Act Amendments) to install sulfur scrubbers to the extent they were in existence in Japan by the end of the 1970s.

This was possible largely because the financial mechanisms employed by Japan—such as subsidization, low interest loans, and tax breaks—often compensate industries for their compliance costs. As a result, firms effectively recover their costs and avoid a decline in market share. Japanese pollution control creates new markets and spurs technological invention, contributing to the enhancement of national environmental quality, industry–government cooperation, and international competitiveness. One example can be found in Japan's development of desulfurization technology. Desulfurization technology first was initiated by the government because of the high level of technological risk involved. Once proven, it was then adopted by firms and incorporated as a national policy mechanism. On the other hand, denitrification technology development was led by private companies. The market value of denitrification equipment was anticipated to be very high, encouraging firms to invest in the technology. These firms were thus motivated to meet regulatory requirements, since this would create a dynamic of expanding market demand (Fujii 2002).

Japanese environmental policy is not enforced punitively, as is often the case in the United States and Europe. Rather, Japanese policy encourages firms to follow central administrative guidance exercised through workable compliance mechanisms. As a result, firms tend to follow administrative guidance enabling them to avoid expensive and lengthy legal battles. Much of this is possible because the government maintains an environment of negotiability with industry, granting firms options based upon mutual consensus. When a new environmental policy is adopted in Japan, an in-depth, informal negotiation process precedes the legally required formal negotiation round. Keidanren, being the most prominent association of Japanese businesses, was the first group to carry out this function of negotiating partner on behalf of industry.

While this approach may appear inefficient by neoclassical economic standards, Japanese stakeholders typically find a way to make the policy work. Coordination rather than conflict among stakeholders is one of the various factors that guides Japan to success in pollution control. Voluntary arrangements stress the importance of flexibility in environmental management. A regulatory approach would require standardized performance that may cause a loophole in regulation. A voluntary approach, on the other hand, has been shown to create relief for governmental administrative costs (legal and monitoring) down the road, as well as give flexibility to firms to meet negotiated environmental quality goals. Weidner (1995) also pointed out that a voluntary approach enabled rapid growth by Japanese industry, along with environmental technology development.

Kitamura (2003) summarizes several benefits of voluntary arrangements for industry: 1) voluntary agreements enable industries to appeal to consumers by being considerate toward the environment; 2) the government may give industries favorable treatment in tax and financing for additional investments; and 3) they help companies maintain a good relationship with the public and the government. According to Kitamura, this establishes a horizontal relationship between industry and government, putting both parties on equal footing and enabling productive negotiation. In addition, there is room to institutionalize preferable policy measures and avoid undesirable costs to both government and industry for emerging environmental problems for which no concrete legal or regulatory framework has taken shape.

Development of the Keidanren Voluntary Action Plan on the Environment

The Keidanren Voluntary Action Plan on the Environment was declared in June 1997, just before the Third Conference of the Parties (COP3) under the United Nations Framework Convention on Climate Change held in Kyoto, Japan. Environmental issues already had become prominent concerns within the business community, and Keidanren addressed them in its Global Environmental Charter in 1991. The charter states, "grappling with environmental problems is essential to corporate existence and activities," thus identifying environmental responsibility as a basic principle of business behavior. In 1996,

Keidanren announced its Appeal on the Environment and encouraged industry leaders to map out their own voluntary action programs. These programs focused on the concrete challenges for industry in global climate change mitigation. The Keidanren Voluntary Action Plan on the Environment was submitted the following year based on this appeal.

The plan was made mainly on Keidanren's own initiative, making it automatically non-binding. The development of the plan, however, included subtle language implying its true nature: a "soft-binding" agreement. The Japanese government also conducted talks with Keidanren members on the potential of reducing greenhouse gases before a numerical target was discussed at COP3. At that time, conclusion of the Kyoto Protocol was uncertain. Tanigawa (2004) points out that the Keidanren Voluntary Action Plan was founded upon the strategic decisionmaking of industries, solidifying their commitment to the plan. It helped industries demonstrate their cooperation and, at the same time, allowed industries to avoid incurring a further burden in case the government failed to control emissions from both the residential and transportation sectors.

After the Kyoto Protocol had been negotiated, an additional approach to the Japanese voluntary arrangements, called "step-by step," was adopted. This approach created space for additional instruments to be introduced if domestic efforts turned out to be insufficient at meeting environmental quality targets. The implications for industry from the step-by-step policy additions included a regulatory threat to the effect that additional instruments, such as an environmental tax or mandatory cap-and-trade, might be introduced if firms failed to meet environmental quality targets.

Origins of the Plan

In 1997, when the plan was first announced, 38 sectors signed on and submitted their own targets, with actual numerical values. Participants consisted of not only manufacturing and energy-converting sectors but also retail and wholesale, finance, transportation, and trade sectors. The wide range of sectors included makes the Keidanren plan unique in the realm of environmental policy. Much of this was made possible because of Keidanren's status as a powerful and influential business association.

Keidanren stimulates its members to broaden the plan's activities. Currently, 50 business associations, 1 enterprise group, and 7 railway companies are active participants in the plan. The participating sectors together emitted 505 million metric tons of CO_2 in 1990, representing approximately 82% of the total amount of CO_2 from the industrial and energy-converting sectors (615 million metric tons) and 45% of national greenhouse gas (GHG) emissions (1.12 billion metric tons).

The overall target was based upon emissions from industrial and energy-converting sectors. This implies a relatively weak contribution from other sectors, namely the commercial and transportation sectors. Among 58 participants, 35 are from industrial and energy-converting sectors and all of them submit their own follow-up reports to the Keidanren's annual follow-up survey. On the other hand, 3 out of 10 commercial sectors and 7 railway companies do not participate in Keidanren's follow-up process, even though their activities are part of Keidanren's Voluntary Action Plan. In sum, 48 sectors, of which 35 are from industrial and energy sectors, 7 are from commercial and household sectors, 6 are from transportation sectors, joined Keidanren's periodic follow-up process (Table 3-1). A periodic review drives industries to continuously tackle this issue.

In the plan, individual firms are bound within their respective industrial associations, not their respective enterprise levels (size of their business). This means that the commitments, of which the emissions targets are the most significant, are set at the sector level, not by individual firms. The basic idea of the Keidanren Voluntary Action Plan is equivalent to a unilateral commitment by industries, whereby industries make commitments of their own free will. Therefore, individual firms have no legally binding targets; as noted by the plan language, "they will endeavor." However, this is recognized as a commitment by industries to manage emissions from industrial sectors. These industry commitments then gain recognition within the governmental climate plan by attaining their Kyoto target. In this context, implementation reports are required to be submitted and examined by relevant governmental bodies, such as the Advisory Committee on Natural Resources and Energy and the Industrial Structure Council.

The Keidanren Voluntary Action Plan has no economic incentives. Nevertheless, many member organizations and corporations responded to the Keidanren and became active participants. This behavior is explained rationally with the traditional relationship between Keidanren and its member companies. Being a representative of business interests, the Keidanren negotiates various policies that affect business activities with the government. Each company benefits from Keidanren membership by being a part of a community. The interdependence of Keidanren and its member companies has existed over many years.[1] Member companies regard this long-term relationship with Keidanren as important and very often support its activities, even if it reduces the companies' profits in the short term.

Objective of the Plan

The Keidanren committed to stabilizing its greenhouse gas emissions at 1990 levels in absolute terms by 2010. In addition to the overall target, each participant has set its own targets, which vary depending on the characteristics of each industry. Some firms select absolute targets, such as reduction in energy consumption or total CO_2 emissions, and others chose intensity targets, such

TABLE 3-1. Number of Sectors Participating in the FY 2005 Review

	Type of target				
	CO$_2$ emissions		Energy consumption		
	Quantity	Intensity	Quantity	Intensity	Total
Industrial and energy converting sectors	12	9	5	15	35
Offices and household sector	2			3	7
Transportation sector	2	3		1	6
Total	16	12	5	19	48

Source: Nippon Keidanren (2005).

as energy consumption or CO$_2$ emissions per unit of output. Improving the energy efficiency of products also can be a target. Of the 48 sectors that submitted fiscal 2005 follow-up reports, 16 defined their goals in terms of the reduction of CO$_2$ emissions, 5 defined their goals in terms of a reduction of energy consumption, and 12 and 19 defined their goals in terms of CO$_2$ emissions intensity and energy consumption intensity, respectively[2] (Table 3-1).

Individual industries determined their own targets, after carefully considering their potential based on technical and economic analysis of energy-saving technologies. The key question is whether these targets require any additional efforts. It is difficult to differentiate targeted emissions from what firms could achieve in the absence of the voluntary plan.

When drawing up a new projection of energy supply or industrial science and technology, which is directly reflected in industrial policymaking, the Japanese government often asks industry representatives for their input. For example, a long-term energy plan is reviewed regularly by the government to secure energy supply. Power producers then commit to this process. Their investment in generating capacity, therefore, is required to be consistent with the current long-term energy plan. In the example of nuclear power generation, this means that a baseline for the power sector had been provided by the plan, while power producers drafted their voluntary action plan. As a result, power producers had to set high targets, since the plan was optimistic about expanding nuclear capacity and the number of new nuclear power plants scheduled turned out to be infeasible.

Measures

Many industries identify improvements in energy efficiency as the highest priority. Each industry spelled out concrete measures for wider dissemination of existing energy-saving technology, the improvement of equipment, and further technological innovation. Examples of measures to increase efficiency in production processes are: 1) re-utilization of waste heat; 2) waste generation;

3) cogeneration; 4) introducing renewable energy resources; 5) fuel switching; and 6) increased use of nuclear power.

While emissions from the industrial and energy-converting industries are steadily heading for a reduction, emissions from the transportation, commercial, and residential sectors have increased by 20–30% in 2003 compared to 1990. Therefore, efforts in those sectors were considered in the Keidanren Voluntary Action Plan. Currently, 23 industrial associations and companies from those sectors are working to take actions based on their own voluntary programs. In addition, many of the industrial and energy-converting sectors also focus on improving energy conservation in offices and transportation processes.

There are two distinct channels available to industry to contribute emissions reductions from transportation, offices, and household sectors. The first channel is the reduction of industries' own emissions through a rigorous approach to energy-saving management in offices, a shift to more efficient office equipment, and environmental education for employees. The second is the manufacture of energy-efficient products by appliance makers, which results in end-use energy conservation. The development and diffusion of various kinds of new, high-efficiency products needs to be encouraged, together with the provision of information to consumers. Energy Service Company operations also promote the commercialization of energy management systems.

In addition to domestic efforts, overseas activities that help reduce CO_2 emissions, such as reforestation, also are considered important measures. The Clean Development Mechanism (CDM) and Joint Implementation (JI) projects specified under the Kyoto Protocol are recognized as complementary measures for business to compensate their sector shortages. Many private companies are actively involved in these activities, as well as voluntary participation in some carbon funds, such as the World Bank's Prototype Carbon Fund (PCF) and the Japan Greenhouse Gas Reduction Fund, a fund established by a collaboration of private companies in Japan. Table 3-2 summarizes JI/CDM projects in which Japanese private companies are involved. The World Bank estimated that Japanese entities (mostly private) purchased approximately 20% of emissions reductions in the world market (The Word Bank 2005).

Reporting

The Keidanren Voluntary Action Plan is recognized as a key instrument to bring industrial emissions under control and achieve Kyoto targets. As a result, the government monitors its achievements and potential to meet the Kyoto target as a whole. While Keidanren annually conducts self follow-up surveys, which it makes public, a joint committee has been established in the Advisory Committee on Natural Resources and Energy and the Industrial Structure Council to conduct an annual review of the follow-up surveys submitted by industries.

TABLE 3-2. Joint Implementation (JI)/Clean Development Mechanism (CDM) Projects Approved by the Japanese Government

JI / CDM	Project proponents	Host country	Project title	Reductions 1,000 tons	Concerned ministries
JI	NEDO	Kazakhstan	Model Project for High-Efficiency Gas Turbine Technology	62	METI
CDM	Toyota Tsusho	Brazil	V&M Tubes do Brasil Euel Switching Project	1,130	METI
CDM	J-Power	Thailand	Yale Rubber Wood Residue Biomass Project	60	METI, MAFF
CDM	INEOS	Korea	HFC Decomposition Project in Ulsan	1,400	MEIT, MOE
CDM	KEPCO	Bhutan	e7 Bhutan Micro Hydro Power CDM Project	0.524	METI
CDM	Japan-Vietnam Petroleum	Vietnam	Rang Dong Oil Field Associated Gas Recovery and Utilization Project	677	METI
CDM	Sumitomo Corp.	India	Project for GHG Emission Reduction by thermal oxidation of HFC 23 in Gujarat, India	3,380	MEIT MOE
CDM	Chubu Electric Power	Thailand	A.T. Biopower Rice Husk Power Project in Pichit	84	MEIT MOE
CDM	J-Power	Chile	Graneros Plant Fuel Switching Project	14	METI
CDM	TEPCO	Chile	Methane Capture and Combustion from Swine Manure Treatment for Peralillo	79	METI
CDM	TEPCO	Chile	Methane Capture and Combustion from Swine Manure Treatment for Corneche and Los Guindos	84	METI
CDM	TEPCO	Chile	Methane Capture and Combustion from Swine Manure Treatment for Pocillas and La Estrella	247	METI
CDM	Showa Shell KK	Brazil	Salvador da Bahia Landfill Gas Management Project	665	METI
CDM	NEDO	Vietnam	Model Project for Renovation to Increase the Efficient Use of Energy in Brewery	10	METI
CDM	Kashima Corp.	Malaysia	Krubong Melaka LFG Collection and Energy Recovery CDM Project	60	METI MLIT
CDM	Shimizu Corp.	Armenia	Nubarashen Landfill Gas Capture and Power Generation Project in Yerevan	135	MOE METI, MLIT
CDM	Showa Shell KK	Brazil	Irani Biomass Power Generation Project	180	METI

Source. Kyoto Mechanisms Information Platform (2005).

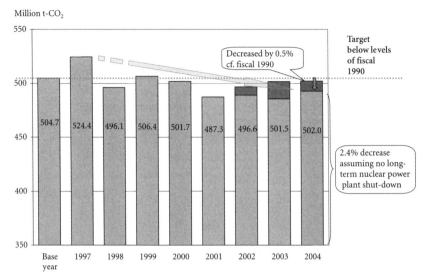

FIGURE 3-1. CO_2 Emissions from 35 Industries in the Industrial and
Energy-Converting Sectors
Source: Nippon Keidanren (2005).

Furthermore, Keidanren established an Evaluation Committee to make business
efforts more transparent and credible. The committee investigates industries'
follow-up reports and provides feedback, most of which is incorporated into
individual industries' activities and reports the following year. This forms a well-
designed Plan-Do-Check-Act (PDCA) cycle for the program.

Accomplishments of the Plan: Participant Perspectives

This section reviews accomplishments of the plan from the perspective of participants. Each participant annually reports its activity and the Keidanren
reports a general overview of participants' activities.

Self Follow-up by the Keidanren

The 2005 follow-up to the Keidanren Voluntary Action Plan on the Environment reports that the 35 business associations in the industrial and
energy-converting sectors that participated in the survey emitted 502 million
metric tons of CO_2. This emissions level is slightly below base-year emissions
(Figure 3-1, Nippon Keidanren 2005). In the years from 2002 to 2004, some
nuclear power plants shut down due to safety problems. Thus, a decrease in
nuclear power generation was compensated for by oil-fired and coal-fired
power generation and this resulted in a worsening of emissions intensity. The
Federation of Electric Power Companies (FEPC) estimates that long-term

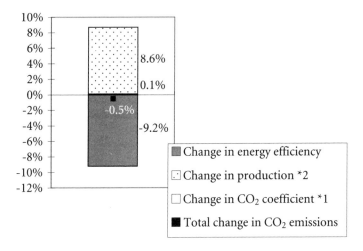

FIGURE 3-2. Factor Decomposition of CO_2 Emissions Change between
FY 1990 and FY 2004
Notes: CO_2/MJ for fuel use; CO_2/kWh for electricity consumption.
Source: Nippon Keidanren (2005).

nuclear power shut-down caused a degradation of emissions intensity by 41 g
CO_2 per kWh generation; therefore, the overall emissions from the 35 sectors
would have been approximately 493 million metric ton CO_2—2.4% below the
base-year emissions—if nuclear power capacity was at the historic high.

The Keidanren reports that the industry and energy sectors are on track.
Among 35 sectors, 20 reported a reduction in CO_2 emissions compared to fiscal 1990, 11 of which have set their targets in terms of reductions of CO_2
emissions. Of the 20 sectors that reported reductions in terms of energy consumption, 5 have set their targets in terms of energy consumption. Of the 20
sectors that focused on intensity targets, 14 reported improvements in their
indices compared to 1990.

The Keidanren evaluates business efforts in two ways. The first type of evaluation is factor decomposition of CO_2 decrease into change in CO_2 coefficient
(CO_2/MJ for fuel use, and CO_2/kWh for electricity consumption), production
change and energy efficiency (energy use per unit production).[3] The Keidanren reported 0.5 percent decrease is decomposed into 0.1% increase caused by
changes in CO_2 coefficient, 8.6% increase caused by changes in production, and
9.2% decrease caused by changes in energy efficiency, which are, in large measure, the result of business efforts (Figure 3-2).

The second type of evaluation is an international comparison of energy efficiency. This was recommended by the evaluation committee, which pointed out
that it is important to assess Japanese companies' performance against an international standard. According to the international comparison of fossil power

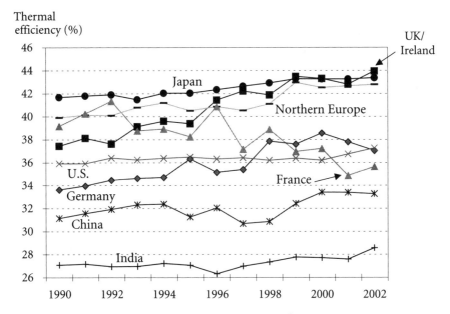

FIGURE 3-3. Comparison of Fossil-Power Plant Efficiency[4] in Japan with Other Countries
Source: FEPC (2005).

plant efficiency conducted by FEPC, Japan already has achieved the highest level of efficiency since 1990. Figure 3-3 indicates a steady improvement in fossil fuel efficiency in Japan after 1997, when the Keidanren Voluntary Action Plan started.

The Keidanren concluded that in total there is steady progress in CO_2 intensity improvement or energy consumption intensity and shifts of energy sources toward those of less carbon content as a result of voluntary efforts by industries. Keidanren also estimates that the business-as-usual CO_2 emissions in fiscal 2010, assuming that the Voluntary Action Plan was not executed from fiscal 2005 on, would increase approximately 14 million metric tons compared to 1990. This means that the goal of controlling emissions below fiscal 1990 levels requires additional business efforts.

Efforts in Each Sector

This section highlights some of the countermeasures taken by individual industries and their outcomes. First, we discuss the iron and steel industry's emissions reductions from production activities. In addition to enhanced energy conservation in factories and offices, the iron and steel industry further reduced emissions through the utilization of plastic waste. The second example is the power sector, which faces great difficulty in meeting voluntary targets. Leaders from this sector announced that they will acquire Kyoto credits to compensate for the shortfall in the event that corporate efforts are not sufficient.

The Iron and Steel Industry

When the Japan Iron and Steel Federation established a plan called the Voluntary Action Program for Environmental Protection by Steelmakers, a numerical target was set to reduce energy consumption by 10% for the year 2010 compared to 1990 levels. Measures to be implemented consist of the following five major approaches: 1) energy saving in iron- and steel-making processes; 2) effective utilization of plastic and other waste materials; 3) energy savings through steel products and by-products; 4) energy savings through international technical cooperation; and 5) utilization of waste energy in areas around steelworks. The industry also committed to achieving another 1.5% reduction through additional measures such as the effective use of plastic waste in blast furnaces and coke ovens. As of the end of 2005, a total of 68 companies were participating in the program, which accounted for 97.4% of annual energy consumption within the industry.

The iron and steel industry consumes large volumes of coal in the iron-making process, in which CO_2 emissions are generated. There are no substitutes for this process; therefore, reducing emissions from this process faces technological difficulties.[5] In addition to this, the emissions coefficients for the iron and steel industries are not clearly defined or available, whereas energy consumption data is available from existing statistics. These are the major reasons why energy consumption was chosen as a target indicator. Thus, the absolute target is preferred to the intensity target since it is regarded as consistent with the Kyoto target.

Actual energy consumption in 1990 was 2,479 PJ. This amount declined to 2,371 PJ in 2004, a reduction of 4.4% compared to 1990. CO_2 emissions attributed to energy consumption also decreased by 10 metric tons from the 1990 level, equivalent to a 5.1% reduction. It is estimated that the industry's effort itself reduced emissions by 14 million metric tons, whereas external factors, such as electricity intensity and other economic factors, caused negative effects amounting to a 4-million metric ton increase.

The iron and steel industry committed to an additional effort to reduce another 1.5% of its energy consumption through the utilization of plastic waste in the production process. Plastic waste materials from industry have been used since 1997 as reducing agents either in blast furnaces or coke ovens. The use of general plastic waste has accelerated since April 2000, when the Container Packaging Recycling Law went into force. In 2004, the use of plastic waste materials in the iron and steel industry reached 410,000 metric tons and is estimated to reach 460,000 metric tons in 2005 (Figure 3-4). The iron and steel

10,000 tons

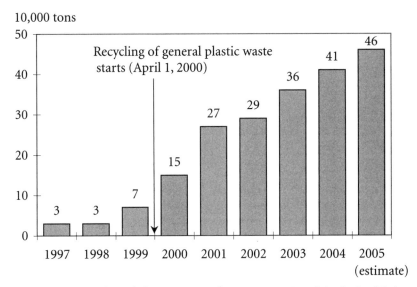

FIGURE 3-4. Recycling of Plastic Waste and Its Treatment Result in the Steel Industry[6]
Source: JISF (2005).

industry is aiming to reinforce its ability to treat plastic waste to reach a capacity of one million metric tons for 2010.

The Power Sector

The Federation of Electric Power Companies measures its CO_2 emissions per kWh of energy use by the end user (CO_2 emissions intensity). The power sector's target is to reduce emissions intensity by approximately 20% from 1990 levels. The target year was set as fiscal 2010. The amount of CO_2 emissions related to the use of electricity is obtained from total electric power consumption multiplied by the CO_2 emissions intensity. From these values, electric power consumption fluctuates, as it is affected by various factors, such as weather and business circumstances, which are beyond power producers' control. The intensity target is selected as it reflects power producer efforts more clearly, although it is also affected by fluctuation of energy demand.[7] The 20% reduction target was based on government projections of long-term energy supply and demand, together with the development plan of nuclear power in 1996, when the plan was first drafted. This then became an extremely challenging target due to substantial delays in the development of nuclear power plants.[8]

The CO_2 emissions per unit of end-user electricity (electricity intensity at user end) was 421g CO_2/kWh in 1990, and the target has been set as 340 g CO_2/kWh in 2010. The actual figure in 2001 was 379 g CO_2/kWh—an improvement of approximately 10% compared to the 1990 level. From 2002 to 2004, however, emissions levels worsened due to the long-term shut-down of nuclear power plants and compensation for the loss of generating capacity by oil- and coal-fueled fossil power generation. The impact of the nuclear power plant shutdowns was greatest in 2003. The nuclear capacity factor gradually rose in fiscal year 2004 as some of these plants came back online. Electric utilities worked to reduce CO_2 emissions through the use of non-fossil energy sources and by increasing the efficiency of power plants. As seen in Figure 3-3, the efficiency of Japan's fossil power plants is the highest in the world as a result of efforts such as raising the combustion temperature in LNG combined-cycle power generation, raising the temperature and pressure of steam in boilers and turbines, and other thermal efficiency management.

Despite the power sector's efforts with both supply-side and demand-side initiatives, it is estimated that CO_2 intensity in 2010 may only reach a 15% reduction, 5% short of the target. There are many factors making it exceedingly difficult to meet the target, such as the fact that plans to develop new generating facilities have been revised due to power sector reforms and stagnating growth in power demand, and, especially, delays in locating sites for nuclear power plants. Nonetheless, the electric utility industry continues to pursue its target with existing measures and by the following goals: 1) enhanced utilization of nuclear power generation; 2) further increase in the efficiency of fossil power production; and 3) project mechanisms such as CDM and JI. Additional measures, including more utilization of the Kyoto mechanisms, will be necessary to meet targets. Some member companies already have implemented overseas projects to reduce or absorb CO_2 and are voluntarily participating in carbon funds, such as the World Bank's PCF (Table 3-3).

Summary

In spite of the increase in production resulting from economic recovery in Japan, as well as increases in CO_2 emissions from electric power generation caused by nuclear power plant shutdowns, CO_2 emissions in fiscal 2004 were slightly below the 1990 level. Although some industries are facing difficulties meeting their targets, broadly speaking business efforts are getting considerable results in terms of energy-saving and CO_2 emissions intensity improvements. Industries are firmly committed to pursuing energy savings in their business activities through the rigorous implementation of voluntary programs. Some industries are forced to make a difficult choice and employ the Kyoto mechanisms to compensate for shortages. This means that their targets are nominally voluntary but well-enforced in practice owing to the design of the program. The self-checking function is further strengthened through the examination of program accomplishments conducted by relevant governmental bodies such as the Advisory Committee on Natural Resources and Energy and the Industrial Structure Council.

TABLE 3-3. Power Sector's Contribution in Various Carbon Funds

Name of fund	Operator	Total contribution
PCF + CDCF + BioCF	The World Bank	$60.5 million
JGRF	JBIC + DBJ	$52 million
GG-CAP	Natsource	$22 million

Source: FEPC (2005).

Accomplishments of the Plan: External Analysis

This section reviews external analysis of the Keidanren Voluntary Action Plan. Recognition of the Keidanren plan is mixed. Industries argue that this mechanism delivers cost-effective measures, since industries, which have the best expert knowledge about their business, are self-motivated and flexibility is secured. On the other hand, the majority of environmental nongovernmental organizations are skeptical about the program. They argue that the effectiveness of the plan is difficult to evaluate because the process is not adequately transparent. Another criticism of the plan regards uncertainty in meeting the Kyoto target, which relies on an 8.3% reduction in emissions from 1990 levels from the industrial sector.[9]

Evaluation Committee

The evaluation committee for the Voluntary Action Plan on the Environment was established by the Keidanren in 2002 to include a third-party review function and to reinforce the PDCA quality. The committee is composed of academic experts and works to: 1) assess whether the processes for the collection, aggregation, and reporting of data in the follow-up surveys by participating industries is implemented properly; and 2) make recommendations concerning aspects of the system that should be improved to increase transparency and credibility.

　　Thus far the review process has been conducted once each year in 2002, 2003, and 2004. An evaluation report was released to the public for each of these years (Evaluation Committee for the Voluntary Action Plan on the Environment 2005). In these reports, the following matters were pointed out, and most of them were taken into account by each industry in the next year's follow-up survey.

1. Boundary adjustment: The Evaluation Committee pointed out there is no consistency among producers for an activity boundary to be covered by each industry. Some cover primary products only, while others cover primary and by-products as well. Therefore, the possibility exists that emissions are double counted. On-site power generation by steelworks has thus been

adjusted between power sectors and the steel and iron sectors to avoid double counting. The committee recommended that similar adjustment among sectors should be addressed in accordance with industry-specific reports of all the follow-up survey results. It also recommended that companies covered by the follow-up survey should be limited to those that can continuously submit data and that data at the sector level should include integration of selected individual data to ensure consistency over time.

2. Standardization of assumptions used in the target-year's emissions forecasts: The committee recommended that forecasts for emissions with and without proposed activities (business-as-usual scenario) should be based on the same assumptions across sectors and these assumptions, such as the value and volume of production and other economic circumstances, should be open to public scrutiny. Also, transitional data from planned year to the target year need to be reported in addition to the projection for the target year (fiscal 2010).

3. Reasons for the selection of target indicators: Selection of the target indicators is by participants' choice and there are several varieties of indicators, including: 1) absolute volume of CO_2 emissions target; 2) CO_2 emissions intensity target; 3) absolute value of energy consumption target; and 4) energy intensity target. Sometimes it is difficult for those outside sectors to understand why targets vary across sectors, which may cause opacity. Therefore, the committee noted that each sector needs to explain the reasons for the selection of the target indicators and the basis for the calculation of target values.

4. Factor analysis of emissions changes: To better understand the effectiveness of emissions-reduction strategies and the influence of external factors, the committee recommended making a quantitative examination of changes in: production, overseas shift of production, operation of installations, products mix, and changes in industry structure, with reasons for such changes as far as it is possible to assess.

5. Endeavor to release information concerning intensity results: Absolute targets are consistent with the national target and easy to measure in terms of meeting the Kyoto target. However, the absolute value of emissions is associated with various factors, as mentioned above. Therefore, the committee recommended that outcomes pursued by each individual industry need to be examined by intensity results, in which external factors can be distinguished. The committee encouraged further endeavors to release information concerning intensity results in addition to current follow-up surveys.

6. International comparison of energy efficiency in each industry: The committee emphasized the importance of international comparisons of energy efficiency from the viewpoint of international competitiveness. According to the committee, evaluations such as whether energy efficiency is equivalent to other countries within the sector and how much abatement costs may be

are the basis for managing both economy and environmental issues in an impartial manner.

7. Report on the state of specific efforts being made in business use and transportation sectors: More participation from the business, residential, and transportation sectors is desirable to increase awareness of outstanding increases in emissions and for stressing the necessity for measures in these sectors. The committee noted that industries also need to reduce emissions from their energy use in offices and transportation, that rational target settings are required, and that achievements need to be measured in a quantitative manner.

8. Evaluation of the effects of emissions reductions over product lifecycles: The committee emphasized the necessity of evaluating the effects of emissions reductions over the entire product lifecycle in as uniform and quantitative a manner as possible. This requires evaluation of not only emissions in the manufacturing process but also those when products are put into use in the household and transportation sectors.

The Voluntary Agreement Study Panel

The Voluntary Agreement Study Panel (2001) reported some criticisms of the Keidanren plan as follows:

1. Target setting: The Keidanren set a target to keep CO_2 emissions from industry as a whole below the 1990 level. However, targets for individual industries have been set independently. The process of target setting in each industry is less transparent; neither target nor performance data for each company is open to the public. Therefore, external investigation of whether individual industry targets will be achieved, based on building-up of individual company performance, is difficult.

2. Program coverage: The Keidanren plan includes only large-scale enterprises that are members of the Keidanren; therefore, small- and medium-sized enterprises are not included. A mechanism that encourages those nonparticipants to enter is needed to increase program coverage.[10] So far, there is no incentive or disincentive mechanism to stimulate commitment.

3. External sensibility: According to the analysis conducted by the panel, reductions of CO_2 emissions from manufacturing industries for 1990 to 1999 were highly dependent on the improvement of emissions intensity in the power sector. In some industries, such as iron and steel and ceramic, stone and clay, substantial decreases of CO_2 emissions arose from changes in production. In other words, emissions from these sectors are sensitive to external factors. According to the panel, the net decreases attributable to industrial efforts are not clearly defined.

4. Mechanisms to ensure compliance: The Keidanren plan is a unilateral, voluntary commitment of industry; therefore, no administrative organ is

explicitly engaged in compelling compliance. Although relevant governmental bodies review accomplishments of the program annually, the panel
asserted that this is not enough. Information is insufficiently disclosed for a
third party attempting to trace an entire follow-up process. In addition, the
current system doesn't assign responsibility. Therefore, role-sharing among
government, industrial associations, and individual companies, especially in
a negative situation, is obscure.

Summary

One interesting question is whether the plan had an impact on emissions compared to the baseline that would have occurred in the absence of the plan.
Ironically, the wide coverage of the plan—most key industries in Japan are
involved—makes it methodologically impossible to establish a comparison
group approach, as is applied in other chapters of this book. The ambitiousness
of company reduction targets is difficult to measure. However, the robustness
of targets is ensured to a certain extent because they are consistent with the government's long-term projections. In addition, the PDCA quality of the plan is
secured through government review. None of the other measures within the
governmental plan to attain its Kyoto target are periodically evaluated to the
same depth. The transparency and credibility of the Keidanren plan, therefore,
are relatively reliable compared to other measures that have no review process.

Conclusion

Since the Keidanren Voluntary Action Plan was submitted in 1997, major
industry associations and companies from various sectors have signed on and
endeavored to carry out the program. Whereas voluntary programs are replaced
by mandatory schemes such as taxes or carbon cap-and-trading schemes in
most European countries, voluntary programs are used more extensively and
effectively in Japan. The Keidanren self follow-up surveys demonstrate that
there are substantial positive outcomes of the program. In fact, emissions from
participating sectors show a broad, declining trend after 1997 (see Figure 3-1).
In addition, without the plan, business-as-usual emissions from participating
industries would increase 14 million metric tons in fiscal year 2010 compared
to the base year. A large part of the success of the Keidanren Voluntary Action
Plan is based on distinctive features of Japanese industrial policies that are
summarized below.

First, the Keidanren played a major role when the plan was formed. The
Keidanren is an all-encompassing business association that has a powerful
influence on both government and business. For businesses, cooperation with
the Keidanren is a rational solution since while they may potentially forego

some profit in the short run, the costs of exclusion from the Keidanren are much higher in the long run.

Second, there are clear goals, based on a long-term national energy plan, that have been developed and negotiated through in-depth discussions among administrative officials, experts, and industries. One advantage of the voluntary approach in Japan is that there are multi-level interactions between the voluntary program and other policies, not only for environmental purposes but also for policies on energy conservation, energy security, and structural change of the economy. The goals are harmonized in a comprehensive picture to achieve a nearly socially optimal plan.

Finally, the plan is not truly voluntary despite its name. There are several factors that compel companies to comply with the plan, as follows: 1) the cooperative relationship between Keidanren and companies; 2) the threat of other instruments, such as a tax or cap-and-trade scheme, if targets are not met; and 3) awareness of private companies' social responsibility. In fact, many companies report that they are trying to acquire Kyoto credits to meet their voluntary target in case their corporate efforts are not enough. Some of these factors depend on the particular characteristics of the government–industry relationships in Japan.

References

Evaluation Committee for the Voluntary Action Plan on the Environment. 2005. Voluntary Action Plan Evaluation Report for fiscal 2004 follow-up surveys (in Japanese). Tokyo, Japan: Nippon Keidanren.

FEPC (Federation of Electric Power Companies of Japan). 2005. Environmental Action Plan by the Japanese Electric Utility Industry. Tokyo, Japan: FEPC.

Fujii, Yoshifumi. 2002. Development of Pollution Control Technology and Industrial Organization: Dynamic Interactions between Regulatory Policy and Pipe-end Technology Development in Japan. In *Development and Environment: The Experience of Japan and Problems in East Asia* (in Japanese), edited by Tadayoshi Terao and Kenji Otsuka. Chiba, Japan: Institute of Developing Economies, 79–106.

Imura, Hideaki, and Rie Watanabe. 2003. *Voluntary Approaches: Two Japanese Cases*. Paris, France: Organisation for Economic Co-operation and Development.

Investigation Committee on Japan's Air Pollution Experience. 1997. *Japan's Experience in the Battle against Air Pollution: Working towards Sustainable Development* (in Japanese). Tokyo, Japan: The Pollution-related Health Damage Compensation and Prevention Association.

Ito, Yasushi. 2004. The Role of R&D Support Schemes for Environmental Technologies: Developing Desulphurization Technologies in Japan. In *Environmental Policy in a Changing Asia: Industrialization, Democratization, and Globalization* (in Japanese), edited by Tadayoshi Terao and Kenji Otsuka. Chiba, Japan: Institute of Developing Economies, 243–272.

JISF (Japan Iron and Steel Federation). 2005. Ongoing Commitment of the Steel Industry against Global Warming. Tokyo, Japan: JISF.

Kitamura, Yoshinobu. 2003. *Local Environmental Law and Policy* (3rd ed.) (in Japanese). Tokyo, Japan: Dai Ichi Hoki.

Kyoto Mechanisms Information Platform. 2005. Information on Japan. Tokyo, Japan: Overseas Environmental Cooperation Center.

Nippon Keidanren. 1997. Keidanren Voluntary Action Plan on the Environment. Tokyo, Japan: Nippon Keidanren.

Nippon Keidanren. 2005. Results of the Fiscal 2005 follow-up to the Keidanren Voluntary Action Plan on Environment. Tokyo, Japan: Nippon Keidanren.

Nishijima, Yoichi. 1997. *Re-engineering Global Environment* (in Japanese). Tokyo, Japan: Seibun Do.

Tanigawa, Hiroya. 2004. Incentive Structure for Japanese Companies to Respond to Environmental Requirements on Their Own Initiatives (in Japanese). Discussion paper 04-J-30. Tokyo, Japan: Research Institute of Economy, Trade and Industry.

Terao, Tadanori. 1994. Industrial Policy and Industrial Pollution in Japan. In *Development and the Environment: Problems in Rapid-growing Asian Economies* (in Japanese), edited by Kojima Reeitsu and Shinozaki Shigeaki. Chiba, Japan: Institute of Developing Economies, 265–347.

Voluntary Agreement Study Panel. 2001. Report on Voluntary Agreement (in Japanese). Tokyo, Japan: Commercial Law Center.

Weidner, Helmut. 1995. Reduction in SO_2 and NO_2 emissions from Stationary Sources in Japan. In *Successful Environmental Policy*, edited by Martin Jänicke and Helmut Weidner. Berlin, Germany: Sigma, 146–172.

World Bank. 2005. State and Trends of the Carbon Market 2005. Washington DC: World Bank.

Notes

1. The Keidanren's former organization was established just after World War II to rebuild the Japanese economy. Since then, it has helped the Japanese economy to develop. In May 2002, it became a comprehensive economic organization by amalgamation with the Japan Federation of Employers' Associations, called the Nikkeiren. As of June 21, 2005, its membership is composed of 1,329 companies, including 93 foreign-owned companies, 130 industrial associations, and 47 regional economic organizations.

2. Some industries set more than two target indicators; therefore, the total number exceeds 48.

3. The indices with the closest relation to energy consumption in each industry were selected. The changes in production of the 34 participating industries in the industrial and energy-converting sectors are weighted averages applying the indices of each industry to CO_2 emissions.

4. Fossil efficiency is the gross generating efficiency based on the weighted averages of efficiencies for coal, petroleum, and gas (low heat value standard). Comparisons are made after converting Japanese data (high heating value standard) to low heat value

standard, which is generally used overseas. Low heat value figures are around 5–10% higher than high heat value figures. The number does not cover private power generation facilities.

5. This is true as long as new steel is produced from iron ore. However, utilization of recycled iron can be an alternative to reduce process emissions with production amount of steel and iron unchanged.

6. Utilization of plastic waste has accelerated since 2000, when social systems to recycle plastic waste from households were established.

7. It should be noted that an increase in energy demand will worsen CO_2 intensity in the short term, since a marginal generating facility is, in most cases, a fossil power plant. This is quite different from most industries, where economy of scale works and improves productivity, thus, lets CO_2 factor down. In the long term, however, there is room to improve CO_2 intensity when a more efficient generating facility is established and starts operating.

8. In 1996, more than ten new nuclear power plants were scheduled to open by 2010. The current plan amended the number to only three, which are currently under construction and will be put into service by 2010.

9. Before the Kyoto target achievement plan came out, there was a 7% reduction required for CO_2 emissions from energy use. This number was revised and strengthened to 8.6% when the plan was approved by the cabinet on April 28, 2005.

10. As of today, the Keidanren plan covers 82% of industry emissions in the base year (1990).

4

Climate Change Agreements in the United Kingdom
A Successful Policy Experience?

Matthieu Glachant and Gildas de Muizon[1]

As part of the national strategy to meet its Kyoto obligations, the UK government in 2001 launched 48 Climate Change Agreements (CCAs) that were negotiated with trade associations representing energy-intensive industries. These CCAs are embedded in a sophisticated policy mix that combines an energy tax (the Climate Change Levy) and an emissions trading system. The levy provides the industries with an incentive to enter into CCAs by granting an 80% tax rebate to participating companies, while they can comply with their CCA target by purchasing allowances through the emissions trading scheme (ETS).

One can argue that CCAs look like the ideal voluntary agreement. In particular, they fit well with the recommendations on the design of voluntary agreements found in reference policy documents (e.g., OECD 2004). They have been designed under a clear and credible threat— paying the full energy tax in their absence. Significant efforts were made during the negotiation process to establish the business-as-usual (BAU) trend to measure the genuine strictness of the CCA targets. Monitoring and enforcement provisions also are particularly well specified. Agreements are enforceable contracts that allow for punishment of noncompliant companies. Clear interim and final targets are set. To sum up, CCAs seem well equipped to be efficient.

Have CCAs met these expectations? In particular, have they led to more environmental improvements than what would have been observed without CCAs? In this chapter, we describe this policy experience and implement a simple methodology to assess the environmental strictness of CCAs targets. This method exploits the idea that the ETS market price reveals the marginal abatement costs of the CCA participants.

Our conclusion is that CCAs have been successful in meeting energy consumption targets but those targets probably were modest for the majority of companies. Contrary to the expectations based on theoretical considerations, CCAs may not have delivered much more environmental improvement than what would have happened without them.

UK Climate Change Policy

Goals

The United Kingdom has been a strong supporter of the Kyoto Protocol and has played an active role in international negotiations. Accordingly, the UK target under the European burden-sharing agreement exceeds the average EU target. This agreement commits the United Kingdom to keeping annual greenhouse gas emissions during the 2008–2012 period to 12.5% below 1990 levels. Despite this, the United Kingdom is one of the few EU Member States to be on course to meet its Kyoto obligations (if we exclude the specific situation of the new Member States of Central and Eastern Europe). By 2004, UK greenhouse gas emissions had fallen by 14.6% relative to 1990 levels. As shown in Figure 4-1, the United Kingdom has complied with its Kyoto targets since 2002–2003.

To a large extent, the United Kingdom's proactive position on climate change derives from its declining trend in emissions during the 1990s. An important driver of this trend has been large-scale fuel switching, mainly the displacement of coal by gas in electricity generation and manufacturing associated with the privatization and the restructuring of the energy sector. As a result, carbon dioxide emissions declined by 5.6% between 1990 and 2004. Emissions of two major non-CO_2 greenhouse gases, methane and nitrous oxide, have fallen by 50% and 40%, respectively, since 1990. Emissions of fluorinated compounds have been cut by 25% since 1990.

In addition to its Kyoto obligations, the British government has set a national CO_2 goal as part of the UK climate change program published in November 2000. It sets out a far-reaching strategy for reducing emissions and for adapting to the effects of climate change. It contains an integrated package of measures to reduce the emissions of greenhouse gases across all sectors of the economy. It also announced the government's commitment to move beyond its Kyoto Protocol target by reducing the United Kingdom's CO_2 emissions by 20% below 1990 levels by 2010. This target has been presented as an interim step toward an ambitious long-term goal of cutting CO_2 emissions by 60% by 2050. This reflects the United Kingdom's willingness to establish a leadership role in climate change issues in the international arena.

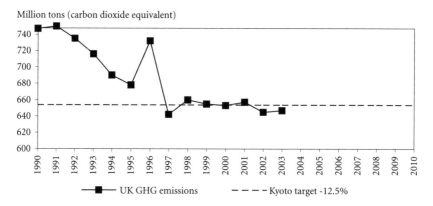

Million tons (carbon dioxide equivalent)

—■— UK GHG emissions   – – – – Kyoto target -12.5%

FIGURE 4-1. United Kingdom Greenhouse Gas (GHG) Emissions and Kyoto Targets

Whereas projections indicate that the United Kingdom probably will meet its Kyoto target, the attainment of the national CO_2 target is unlikely. In 2004, the Department of Trade and Industry forecasted a 14% reduction (DTI 2004), and Tony Blair announced: "We are on track to meet our Kyoto commitment. . . . But we have to do more to achieve our commitment to reduce carbon dioxide emissions by 20% by 2010."

Policy Instruments

The UK program provides a particularly rich and complex example of policy interaction. In particular, an innovative policy mix has been elaborated to reduce the emissions of the industry sector, which accounts for about one-third of total UK emissions (see Figure 4-2).

Following Lord Marshall's conclusions in his report to the Chancellor of the Exchequer (1998), economic instruments play a crucial role in UK policy. The policy includes the Climate Change Levy (CCL), which was launched in 2001. The CCL is a downstream tax on energy used in the industrial, commercial, and public sectors. The levy is revenue neutral: receipts are recycled back to business through both a reduction in employer national insurance contributions and the financing of schemes promoting energy efficiency. This is in line with the double dividend argument, since it shifts the tax burden from labor to CO_2 emissions. The tax covers all types of energy sources except fuel oil, which already is targeted by a preexisting tax. CCL rates vary across energy sources from £5 per ton of CO_2 for coal[2] to £10 for electricity.

In his report, Lord Marshall also recognized the need "for the levy to be designed in a way that delivered worthwhile improvements to energy efficiency and reductions to carbon emissions whilst at the same time safeguarding the competitiveness of UK business and, in particular, the energy intensive users."

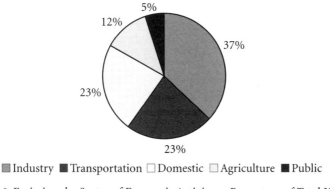

5%

12%

37%

23%

23%

23%

▨ Industry ■ Transportation □ Domestic □ Agriculture ■ Public

FIGURE 4-2. Emissions by Sector of Economic Activity as Percentage of Total UK Emissions

The competitiveness concern clearly has driven the introduction of CCAs, as the government offers an 80% levy reduction to selected industrial sectors in exchange for a quantified, negotiated commitment to reduce energy consumption or CO_2 emissions.

In addition to the levy and the CCAs, the United Kingdom also launched a pilot emissions trading scheme in 2002, three years before the European-wide ETS. Again, this reflects the United Kingdom's willingness to ensure its leadership over climate change issues and to exploit a first-mover advantage.

The UK ETS was not established for CCA companies. The initial project was to set up a cap-and-trade scheme with the voluntary participation of non-CCA companies through the auction of an incentive subsidy to make absolute abatement targets. The program then was extended to allow CCA companies to enter the market on a baseline-and-credit model.

Description of Climate Change Agreements

Sector Coverage

The CCAs consist of quantitative, energy-use reduction objectives to be met by 2010. They were negotiated with 48 sectoral associations covering 44 industrial sectors. Since then, two sectors have left the scheme (reprotech and vehicle builders and repairers). The scheme covers around 12,000 individual sites (5,500 companies) and nearly 44% of total UK industry emissions.

The introduction of the CCAs with a tax exemption initially was aimed at reducing the tax burden for energy-intensive firms. A limited number of industrial activities were then identified by the UK government to be eligible for the scheme: aluminum, cement, ceramics, chemicals, food and drink, foundries, glass, non-ferrous metals, paper, and steel. Lobbying efforts made by specific sectors to benefit from the tax exemption, however, led to a much larger sector coverage—including, for instance, agribusiness and the motor industry. In the end, all eligible sectors negotiated a CCA.

Targets

The contribution of CCAs to British climate change policy is significant. The CCAs are expected to represent more than 25% of CO_2 abatement in industry, which corresponds to a reduction of 9.2 $MtCO_2$ below BAU by 2010 and an average 12% reduction below 2000 levels.

Targets could be expressed in relative terms (per unit of output) or in absolute terms irrespective of production level. More specifically, four types of targets have been proposed: 1) relative energy (GJ primary energy per unit of output, '*Rel E*' in Table 4-1); 2) relative carbon (tons of carbon per unit of output, '*Rel C*' in Table 4-1); 3) absolute energy (GJ primary energy, '*Abs E*' in Table 4-1); or 4) absolute carbon (tons of carbon). Almost all of the sectors have opted for primary energy-use targets for a practical reason: energy consumption already is monitored.

All but two sectors have opted for relative targets. The two exceptions are the aerospace and the steel industries. The advantage of relative objectives for businesses is that they prevent fluctuations of future output levels from affecting compliance efforts. By contrast, absolute targets may be advantageous for sectors anticipating a decrease in production, which might explain the steel and aerospace industries' decisions. The steel industry has faced several operational difficulties and major structural changes since the CCA was adopted, resulting in a sharp decrease of output.

Negotiated targets vary considerably across sectors, as shown in Table 4-1. This reflects the influence of sector-specific variables, such as growth rate, hypotheses on spontaneous technological evolution, market structure, and the negotiating skills of the sector association. The baseline year the targets refer to also is sector-specific due to differences in terms of energy data availability or to the existence of early energy-efficiency actions.

Contractual Arrangements, Monitoring, and Enforcement

Contrary to most negotiated agreements in other EU countries, CCAs include a complete monitoring and enforcement apparatus: CCAs set a final 2010 target and also interim targets for each of the two-year milestones (2002, 2004, 2006, and 2008). For each milestone, individual facilities have to report energy and production data to their sector association. Independent crosschecks can be undertaken by the Department for Environment, Food and Rural Affairs (DEFRA).

TABLE 4-1. Summary of Sector Targets

Sectors	Type	Target by 2010	Base year
Aerospace	Abs E	-8.5%	2001
Agriculture supply	Rel E	-7.1%	1999
Aluminum	Rel C	-32.2%	1990
Brewing	Rel E	-11.6%	1999
Cathode ray tubes	Rel E	-21%	2000
Cement	Rel E	-25.6%	1990
Cementitous slag	Rel E	-10%	1999
Ceramic–fletton	Rel E	-8.1%	2000
Ceramic–materials	Rel E	-10.1%	2000
Ceramic–non fletton	Rel E	-12.4%	2000
Ceramic–refractories	Rel E	-10.33%	2000
Ceramic–white wares	Rel E	-10.22%	2000
Chemicals	Rel E	-18.3%	1998
Craft bakeries	Rel E	-9%	1999
Dairy industry	Rel E	-9.2%	1999
Egg product (BEPA)	Rel E	-9.3%	1999
Egg product (NFU)	Rel E	-11.5%	1995–2000
Food and drink	Rel E	-8.1%	1999
Foundries	Rel E	-11%	1999
Glass	Rel E	-9.2%	2000
Gypsum products	Rel E	-7.2%	1999
Leather	Rel E	-9.8%	2000
Lime	Rel E	-7.9%	1999
Malting	Rel E	-7.8%	1998
Metal forming	Rel E	-7%	1999
Metal packaging	Rel C	-9%	1999
Mineral wool	Rel E	-14.9%	2000
Motor manufacturers	Rel E	-15.3%	1999
Nonferrous	Rel E	-14.7%	1995
Paper	Rel E	-24%	1998
Pigs	Rel E	-16%	1997
Poultry meat processing	Rel E	-12.5%	1995–2000
Poultry meat rearing (BPC)	Rel E	-13.7%	1999
Poultry meat rearing (NFU)	Rel E	-13.7%	1999
Printing	Rel E	-12%	2000
Red meat	Rel E	-10.8%	1999
Rendering	Rel E	-9%	1999
Rubber	Rel E	-10.3%	1999
Semiconductors	Rel E	-59%	2000
Spirits	Rel E	-4.5%	1999
Steel	Abs E	-11.5%	1997
Supermarkets	Abs E	-4.5%	1999
Surface engineering	Rel E	-10.3%	1999
Textiles	Rel E	-9%	1999
Vehicle builders and repairers	Rel E	-10%	2000
Wallcoverings	Abs E	-9%	1999
Wood panel	Rel E	-7.3%	1999

Notes: Rel E = Target expressed in relative energy; Rel C = relative carbon; Abs E = absolute energy; Abs C = absolute carbon.

Three types of contractual arrangements have been available to sectors entering CCAs: 1) one sector-wide agreement with one target and a one-stage all pass/all fail evaluation; 2) an umbrella agreement between the DEFRA and the sector association and underlying agreements between DEFRA and each company; and 3) an umbrella agreement between DEFRA and underlying agreements retained and managed by the sectors.

Most sectors have opted for the second option. In this case, verification involves two steps. First, compliance with the sector target is measured. If it has been met, all target units (i.e., all facilities) are deemed to be compliant and the process stops. If the sector target has not been reached, individual targets are considered. Each facility is recertified only if its individual target is met. If it is not reached, the facility is not recertified, which means it loses its tax exemption for the next two years. It can regain the discount if it complies with the next interim target at the end of the next milestone. However, if a facility fails to comply with its final target in 2010, it will have to pay back all the exemptions it has enjoyed.

The advantage of this two-tiered scheme is that it avoids the cost of individual verification in cases of collective compliance. It is also expected—perhaps naively—that over-achievement by one participant will compensate for the under-achievement of another.

Option three is quite similar to option two and has been chosen by six sectors. Option one was rarely chosen because of free-riding concerns. Under this option, a company meeting its own target can lose its exemption if the whole sector fails to comply.

CCAs and Emissions Trading

Even if the detailed rules governing the interaction between emissions trading and the CCAs had not been developed at the time the CCAs were negotiated, provisions for emissions trading were included in the agreements. The ETS involves two types of participants: the so-called direct participants and the CCA companies. The participation of the direct participants (DPs) is voluntary. The government offered incentive payments to UK companies committing to greenhouse gas emissions reductions. The incentive payments, which amounted to £215 million, were allocated by an Internet auction in March 2002.

The auction was conducted with a descending price clock because it was a procurement auction (or "reverse auction"): the government sought to purchase emissions reductions at minimum cost. The government posted a price per unit of emissions reductions and firms bid on the quantity of emissions reductions that they were prepared to make at that price. In each new round, the government announced a successively lower price and bidders indicated the quantity of emissions reductions that they were prepared to make at the lower price, until the market cleared.

All companies that were not involved in a CCA could participate in the program. A total of 34 qualified for the incentive and shared the incentive payments for accepting a total abatement goal of 4 $MtCO_2$ to 2006. These companies are either large oil companies (BP, Shell) that are not eligible for a CCA since the CCL does not target fuel oil, companies emitting non-CO_2 gases such as HFC (INEOS), or non-energy-intensive enterprises such as banks and retailers.

After taking into account the effect of the yearly abatement profile and corporate tax, the incentive rewards each ton of CO_2 with around £12.

In parallel, CCAs firms can participate in the ETS on a baseline-and-credit basis. If a CCA participant overcomplies with its target, it can receive emissions credits that can be traded on the emissions market. Conversely, a CCA participant can purchase emissions permits on the market to comply with its CCA obligations. As a result, linking the CCAs with emissions trading rewards overachievement and potentially reduces compliance costs.

Programmatic Accomplishments: The Regulator's View

Ex Ante Identification of the Baseline Scenario

The evaluation of the expected environmental impact of any CCA target requires establishing a baseline scenario describing what would have occurred without any policy. This is a difficult task, but significant efforts were made during the negotiation process to provide a benchmark for judging the strictness of the targets.

The energy-consulting firm ETSU provided assistance to DEFRA, comparing the CCA targets with two reference scenarios:

- A baseline BAU scenario that describes what would have occurred if firms did not change their behavior. This scenario basically is an extrapolation of the recent trend.
- An all-cost-effective (ACE) scenario in which all cost-effective measures of energy efficiency are implemented by the companies. In comparison to the BAU scenario, ACE measures include major plant replacement, retrofitting of particular components, and better energy management. It also assumes neither restriction on capital availability nor on managerial time.

ETSU (2001) has estimated that if all cost-effective measures were implemented in the sectors covered by CCAs, it would lead to a reduction of 14.6 Mt CO_2 by 2010. In comparison, the completion of CCAs targets results in a reduction of 9.2 $MtCO_2$ beyond the BAU scenario. As emissions amounted to 100 Mt CO_2 in 2000, this corresponds to a 9% decrease due to CCAs. Presented differently, the agreements would bridge about 60% of the gap between BAU and ACE.

Do these figures indicate strict targets? Basically, the answer depends on the judgment about the strictness of the ACE scenario. On the one hand, it can be viewed as a particular BAU scenario if we assume that companies spontaneously implement cost-saving options. On the other hand, the assumptions of unlimited management time and capital availability are very optimistic. To give a reference point, ETSU has estimated that the price effect of the full-rate levy with no agreements would lead to an abatement of 10.1 $MtCO_2$. Put differently, the CCAs are 10% less environmentally effective than the full tax. Based on this evidence, ETSU drew up a very positive assessment of the negotiation outcomes and estimated that the CCAs' targets would lead to real improvements below the BAU scenario and result in a very modest lowering of the environmental impact in comparison with the full-rate levy.

The Association for the Conservation of Energy[3] disagrees with ETSU's conclusions. Its view is that CCAs targets are very close to BAU. It quotes two different studies, one by the European Commission and the other by the Department of Trade and Industry, which estimate that the BAU scenario would result in a 9% and 13% reduction of CO_2 emissions, respectively, by 2010. Comparing those figures with the CCAs' targets (an average 12% reduction in the period 2000–2010), the conclusions are far less optimistic.

Who is right, the ETSU or the Association for the Conservation of Energy? It is not possible to draw a conclusion without detailed information on the methodology used in the different studies to elaborate the baseline scenarios. This is a common difficulty of ex-ante assessments that have to rely on many assumptions that are always questionable.

Results of the First Two Periods

CCAs specify two-year interim targets. Future Energy Solutions[4] has been mandated by DEFRA to assess the results of the first and second periods (2002, 2004). At the end of the first period (2001–2002), 33 of 48 sectoral associations met their target as a whole. A total of 5,042 sites have been recertified, while 700 have either left the agreements, have not been recertified, or did not submit any data at the end of the milestone period—which implies that their agreements have been terminated. Overall, around 88% of target units have been recertified. It is estimated that the cumulative, absolute energy saving compared to the baseline years is equivalent to 15.8 $MtCO_2$. This figure has to be compared with the overall target, which was 6 $MtCO_2$.

By far, the main contributor to these overachievements has been the steel industry. Due to major changes in this industry (operational difficulties and major structural changes meant that output and CO_2 emissions were reduced significantly), abatement in the steel industry accounted for 9.4 $MtCO_2$ of the 15.8 $MtCO_2$. DEFRA and the steel sector agreed to a target adjustment, which increased from 1.4 $MtCO_2$ to 7.7 $MtCO_2$ for the first period. Taking this adjustment into account, the overachievement of the CCA sector for the first period was estimated at 3.5 MtCO$_2$.

The results of the second period (2003–2004) were even better: 42 out of 46 sectoral associations were recertified. A total of 4,420 target units complied, while 255 have not been recertified. This means that 95% of target units have been recertified. There also was a big overachievement of around 8.9 MtCO$_2$ (5.1 after the steel target adjustment), while the overall target for the second period was 5.5 MtCO$_2$ (9.3 after the adjustment).

Moreover, Future Energy Solutions points out that the CCAs have brought a change in attitude toward energy management in industry. In addition to the savings from the reduced rate of the CCL, they estimated that CCA participants collectively save more than £650 million per year from their reduced energy consumption.

The 2004 Target Renegotiation

Renegotiations of interim targets for CCAs are allowed at the second and fourth checkpoint (2004 and 2008). The 2004 review provided an opportunity for both industry and government to reexamine the assumptions behind the setting of targets and to see if they were too low or too high. Table 4-2 reports the percentage of target adjustment. Unsurprisingly, the better-than-expected performance of CCA participants led to a tightening of the targets of 25 sectors.

In the end, the DEFRA and the CCA companies are very positive about the program's accomplishments. Overall, the emissions reduction is twice as large as expected. In addition, the scheme appears very cost-effective through its connection to the ETS and it has led to very significant behavioral changes in energy management.

Our Ex Post Evaluation

The extent of the success of CCAs in terms of goal attainment may cast some doubt over the genuine environmental strictness of CCAs. Why would profit-making entities abate beyond the targets if abatement was costly? We now turn to our own assessment of the environmental impact of the CCAs. We seek to identify whether they have led to additional greenhouse gas abatement relative to the BAU scenario.

Methodology

From a general point of view, measuring the environmental effectiveness of a voluntary agreement is a difficult task because it requires the identification of a baseline scenario. However, the fact that CCAs are linked with an ETS offers a simple assessment method. It rests on the idea that the prices observed in the

TABLE 4-2. Change in Sectoral Targets Following 2004 Review and Proportion of Units in Each Sector Meeting Their Targets

	2004 review	First period	Second period
Aerospace	1%	100%	100%
Agriculture supply	3%	100%	100%
Aluminum	TBA	100%	100%
Brewing	2%	100%	100%
Cathode ray tubes	3.2%	100%	100%
Cement	0%	100%	100%
Cementitous slag	6.1%	100%	100%
Ceramic–fletton	-11.5%	100%	100%
Ceramic–materials	12.3%	91%	100%
Ceramic–non fletton	0.5%	100%	100%
Ceramic–refractories	0.6%	93%	100%
Ceramic–white wares	6.7%	98%	100%
Chemicals	3.6%	100%	100%
Craft bakeries	18.4%	100%	100%
Dairy industry	2.25%	100%	FAILED
Egg product (BEPA)	TBA	68%	100%
Egg product (NFU)	TBA	99%	100%
Food and drink	2%	100%	FAILED
Foundries	0%	95%	100%
Glass	1%	100%	100%
Gypsum products	0%	100%	100%
Leather	0%	100%	100%
Lime	0%	100%	100%
Malting	0.2%	100%	100%
Metal forming	TBA	100%	100%
Metal packaging	1%	95%	100%
Mineral wool	0%	100%	100%
Motor manufacturers	3%	100%	100%
Nonferrous	TBA	100%	100%
Paper	2.78%	100%	100%
Pigs	TBA	100%	100%
Poultry meat processing	TBA	98%	FAILED
Poultry meat rearing (BPC)	TBA	99%	100%
Poultry meat rearing (NFU)	TBA	83%	100%
Printing	3%	96%	FAILED
Red meat	0%	97%	100%
Rendering	2.5%	100%	100%
Rubber	12.6%	100%	100%
Semiconductors	0%	100%	100%
Spirits	1%	100%	100%
Steel	0.8%	100%	100%
Supermarkets	TBA	100%	100%
Surface engineering	TBA	100%	100%
Textiles	3%	100%	100%
Vehicle builders and repairers	-	21%	CCA TERMINATED
Wallcoverings	TBA	100%	100%
Wood panel	1.87%	100%	100%

Notes: BEPA = British Egg Products Association; BPC = British Poultry Council; NFV = National Farmers Union of England and Wales.

emissions market reveal the abatement costs of the market participants. According to standard microeconomic theory, the price simply equates the marginal abatement cost of any participant if the market is competitive. This result is easily established by contradiction: If one participant's marginal cost was less than the price, he would sell credits or buy credits if their marginal cost was higher. As a consequence, observing non-zero prices means that all participants bear positive abatement costs in equilibrium. Put differently,

Proposition 1. *If the market price is positive, the aggregate abatement delivered by all market participants is higher than BAU abatement. Furthermore, the higher the price, the higher the difference between observed abatement and BAU abatement.*

Before going further, it is necessary to clearly state the limits of this general approach, which are twofold. First, market prices only reflect ex post additionality. And since the actors have set CCA targets several months, or possibly a year, before abatement actions actually have been implemented, ex post additionality may diverge significantly from ex ante additionality. For instance, if a sector experiences a drastic production downturn, meeting absolute targets becomes far less easy than expected and a target that is additional in the ex ante sense might not be ex post additional. We have seen that significant, and probably unexpected, economic fluctuations have occurred in certain sectors (e.g., the steel industry). However, this drawback is mitigated by the fact that most CCA targets are relative.

A second potential problem is related to what BAU abatement means precisely. Proposition 1 assimilates BAU abatement to profitable abatement activities. In practice, companies may not implement profitable abatement actions in a BAU scenario for informational or organizational reasons. If the CCA contributes to removal of these barriers, it leads to additional abatement that is not costly ex post. Our general approach is not able to measure the additionality of profitable abatement, meaning that we tend to underestimate the CCAs' environmental effectiveness.

Having made these general remarks, we come back to the presentation of the details of the methodology. If the CCA companies were the only market participants, Proposition 1 would make the evaluation straightforward. We would directly infer the strictness of the target from the observation of the market price. But there are other types of market participants in the ETS market. In particular, the DPs have been granted CO_2 permits through the auctioning process. In this context, Proposition 1 only implies that the sum of the DPs' allocations and of the CCA participants' targets yield additional abatement if the price is positive. It does not mean that CCA targets alone are additional.

Let us clarify this point using Figure 4-3. In this graph, we adopt the point of view of an individual CCA participant acting in the ETS market. The graph

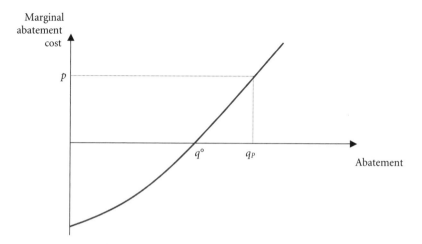

FIGURE 4-3. Marginal Abatement Cost as a Function of Abatement Level

shows its marginal abatement cost as a function of its abatement level denoted q (the dark curved line). Note that the marginal cost is negative below a certain level of abatement $q°$, implying that the polluter selects this level of abatement in the absence of policy. Hence, $q°$ is BAU abatement. In the same graph, the dotted horizontal line shows the market price p. In this context, microeconomic theory predicts that it selects the level of abatement q_P. In market equilibrium, additional abatement beyond BAU is simply the difference $q_P - q°$.

Figure 4-3 shows that an individual CCA company makes additional abatement, just like all other market participants, if the market price is strictly positive. But it does not necessarily mean that its CCA target is additional. To see that, let T denote the target. Then, Figure 4-4 shows what happens for different levels of T. We have three cases:

- If the target T is less than $q°$, the CCA does not have any impact on abatement and the company overcomplies with its CCA target, meaning that it gets credits.
- If the target lies in between $q°$ and q_P, the CCA target yields additional abatement relative to the baseline and we observe overcompliance as well.
- Finally, if the CCA target exceeds q_P, the firm complies with the CCA obligations by abating a quantity q_P and purchasing the additional credits necessary to fill the gap between q_P and T.

So far, we have considered the decision made by a CCA company based on its knowledge of its marginal cost curve. Now, let us adopt the point of view of an outside evaluator who is not informed about individual, marginal abatement costs. He only observes the price p, the initial target T, and the equilibrium abatement q_P. In this informational context, he can only infer with certainty that the CCA target of a company is additional when the firm is a net buyer of

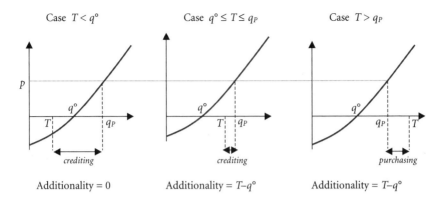

FIGURE 4-4. Different Cases for Marginal Abatement Cost Functions and Target Additionality

credits, that is, if $T < q_P$. By contrast, he cannot conclude when the CCA firm exceeds its target and receives credits. We can only suspect that those receiving a small amount of credits have an additional target so that $q° \leq T \leq q_P$, but without any precise idea of the crediting volume above which the CCA target falls below the BAU level ($T < q°$). To summarize,

> **Proposition 2.** *If a CCA company buys credits, its target exceeds the BAU reduction. By contrast, we cannot conclude if the company is a seller.*

Propositions 1 and 2 provide the basis for evaluating CCA environmental effectiveness in the following. We proceed in two steps. First, we present data on prices to get an idea of the overall environmental strictness of the obligations imposed on both DPs and CCA participants. Then, we analyze the market behavior of CCA participants to identify whether they are net buyers.

ETS Prices

One must be cautious when interpreting UK ETS prices. There is not a unique official source of information. In practice, many traders report transaction prices that are then pooled in a time series by analysts.

Available information shows that during an initial 8-month period, the price increased rapidly to a peak of about £12 in October 2002. Afterwards, the price collapsed very quickly. Since then, it has varied in a narrow range between £2 and £5 (Atkinson 2005). The initial period is most probably an anomaly. At that time, many participants did not have their initial allocation and the volume of transactions was extremely small. In what follows, we rely on the £2–5 range.

The fact that this price is positive unambiguously signals that the overall initial allocation to DPs and CCAs yields additional abatement effort in

TABLE 4-3. Comparison of the UK Emissions Trading Scheme Price with Other Reference Prices

	Price range in £ per ton eCO_2
UK Emissions Trading Scheme	2–5
Other CO_2 markets	
Kyoto MDP credits (Segalen 2005)	1–4
Chicago Climate Exchange market (CCX, 2006)	2–3
EU carbon market (Pointcarbon, 2006)	20
Climate Change Levy rate	
Liquid petroleum gas	3
Coal	5
Gas	8
Electricity	10

comparison with the baseline. But how does this level compare to other reference prices? Table 4-3 reports different benchmarks. To begin with, we compare with the price observed in other CO_2 markets. The UK price appears slightly higher than the price of Clean Development Mechanism (CDM) credits. As CDM crediting is surrounded by much uncertainty, observers generally agree that this price is quite low. The UK ETS price is slightly higher than the price in the Chicago Climate Exchange market as well. This is not surprising since participation in the Chicago Climate Exchange market is purely voluntary. By contrast, the UK ETS price is much less than the EU ETS price. But the EU emissions trading market is probably too recent—it was launched in 2005—to provide a reliable reference point. Recall that the UK prices reached a peak of £12 in the first year.

Table 4-3 also reports the rates of the CCL. Recall that entering into a CCA provides the company with an 80% tax rebate. The initial philosophy was to obtain the same abatement efforts under the CCA while reducing the tax burden. Excluding liquid petroleum gas, which is a limited source of energy in the industry, Table 4-3 shows that CCL rates are higher than the ETS price. This suggests that abatement may not be as large under the CCA.

All in all, UK ETS prices seem quite low. Yet the U.S. acid rain program has shown that there may be a significant gap between (long-term) marginal abatement costs and prices, implying that prices may be an imperfect proxy of additionality. This discrepancy has been observed in the United States because many firms anticipating too high a price made irreversible abatement investments before the price was observed (they installed a lot of scrubbers before the market was launched). Ex post, the use of low-sulfur western coal provided a much cheaper compliance alternative than scrubbers.

We believe that the story is not the same in the United Kingdom because the timing is different. In the United States, the SO_2 market was launched several years after the allowances had been allocated, leaving a lot of time to make investment errors. In the United Kingdom, the market participants also anticipated higher prices, as suggested by the price of the reverse auction (£12) and the market prices observed during the first ten months of the program. But it is doubtful that these wrong anticipations led to inefficient investments, as the CCA targets were set just one year before the launch of the market.

The Market Behavior of CCA Participants

The UK ETS market price is undoubtedly positive. But as previously noted, this does not mean that CCA targets are additional; CCA participants may have been sellers of allowances that, at least in part, arose from a target above BAU (often referred to as "hot air"). In this section, we analyze the market behavior of CCA companies in the ETS market.

Data

We rely on the transaction log of DEFRA, which registers every operation made on each market participant's account. More precisely, it yields:

- The allocation of allowances (crediting). For a CCA firm, it corresponds to the quantity of credits granted to a company overcomplying with its CCA target.
- The retirement of allowances. For a CCA participant, it is the quantity of allowances retired to cover liabilities in respect of their CCA target.
- The quantity of allowances bought and sold.

We complete with information from Future Energy Solutions AEA Technology (2005) on ring-fencing. Ring-fencing occurs when a CCA participant overcomplies with its target. In this case, it has to ring-fence, or protect, its overachievement, otherwise this overachievement is lost and helps the global compliance of the sector association. Once the overachievement has been ring-fenced, it has to be verified to obtain an allocation of permits. The verification can be done at a later period, which enables a participant to add the ring-fenced overachievement over several periods and to make the verification only once.

To make clear these distinctions before entering into the analysis, Figure 4-5 describes the relationship between emissions and the use of allowances. We contrast Case 1, where emissions of the CCA company are less than its target (underachievement), and Case 2, describing the opposite pattern (overachievement).

If a CCA participant does not comply with its target (Case 1), it has two possibilities. It can either buy an amount of permits at least equal to the gap between its emissions level and its target, or decide not to comply, which means that it will be subject to the full CCL for the next period. After having purchased

Case 1: underachievement

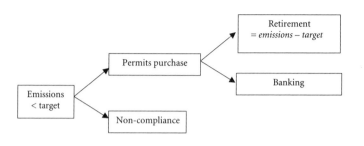

Case 2: overachievement

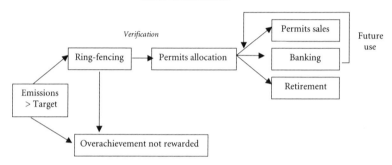

FIGURE 4-5. The Relationship Between Emissions and the Use of Allowances in the United Kingdom

some permits, the CCA participant needs to retire a number of permits equal to its underachievement. If it has purchased more permits than needed for retirement, it can bank them for future use.

On the other hand, an overcompliant CCA participant ring-fences its overachievement. Then, the overachievement has to be verified to obtain an allocation of permits. Finally, allocated permits can be sold on the market, banked for future use, or retired at a later period if the CCA participant needs to use them.

Analysis

Table 4-4 reports the quantity of allowances generated and their use by the different market participants. A striking fact is that only 1,243 CCA companies out of 5,500 have ever traded on the ETS market. The vast majority of the CCA companies have preferred to meet targets through direct action as opposed to trading in the market. We come back to these companies later on.

Table 4-4 shows that the 1,243 CCA companies having made transactions on the market are net buyers of 0.95 $MtCO_2e$. Proposition 2 thus suggests that the CCA scheme is additional in the aggregate for this subset of companies. Table 4-4 also identifies the 1,036 CCA companies (out of the 1,243) that received an allocation less than retirement—meaning that they were obliged to buy allowances to meet their CCA targets. Proposition 2 tells us that their individual target was stricter than the BAU level. Note that they have banked some permits. A possible explanation is that they did so hoping higher permit prices would emerge in the coming years.

By contrast, 207 companies have more permits allocated than retired—allowing them to sell 0.58 $MtCO_2e$ of allowances in the market and to bank 1.91 $MtCO_2e$ for future use. Those are the only companies for which there is the doubt over the strictness of their CCA target.

Table 4-4 also reports data on DPs. While there are only 31, emitting about 30 million tons of CO_2 as compared with 100 million tons by the CCA companies, their overachievements amounted to 7 $MtCO_2$ at the end of 2003. This suggests that the DPs' initial allocation was very generous.

What about the CCA companies that have not made any transactions on the market? A first possible explanation is that they have neither exceeded nor fallen below their CCA target, so they had nothing to buy or sell. In fact, this explanation does not hold—they have ring-fenced a very large quantity of overachievements: 3.20 $MtCO_2e$ for the period 2001–2002 and 5.40 $MtCO_2e$ for the period 2003–2004. These quantities are far larger that the overachievement of companies having traded in the ETS market. In our setting, where we assume that enterprises are profit-making entities unwilling to undertake costly abatement beyond their target, this suggests that their targets were not additional.

Why haven't they converted these overachievements into allowances they could have sold at a positive price? The explanation most probably lies in the verification cost borne by the company to convert overachievements to allowances. According to Dairy Energy Savings, it might cost about £1,000 per site. The importance of ring-fencing shows that this cost has created a barrier to entry into the ETS market.

As an aside, the cost of verification only hinders the entry of sellers, since purchasers obviously have no emissions to be verified. Consequently, this transaction cost only reduces the supply of permits in the market. Accordingly, the market price would have been less than £2–5 per ton eCO_2 if transaction costs were zero. This is not without consequence on our assessment of the UK ETS price. The £2–5 price probably overestimates the marginal abatement cost of the CCA companies in equilibrium (including those having not traded using the market).

In summary, 1,036 companies are shown to have targets stricter than BAU abatement. For 207 companies, it is not possible to draw a conclusion, while the

TABLE 4-4. Market Behavior of UK ETS Participants

	Number of firms	Allocation (MtCO$_2$e)	Retirement (MtCO$_2$e)	Net allocation (MtCO$_2$e)	Net sales (MtCO$_2$e)	Banking (MtCO$_2$e)
		(1)	(2)	(1) – (2)	(3)	(1) – (2) – (3)
CCA participants	1,243	2.76	1.35	1.41	-0.95	2.37
CCA with allocation > retirement	207	2.73	0.24	2.49	0.58	1.91
CCA with allocation < retirement	1,036	0.03	1.11	-1.08	-1.53	0.46
CCA period 1: 2001–2002		1.44	0.59	0.85	-0.60	1.32
CCA period 2: 2003–2004		1.32	0.76	0.56	-0.36	0.92
Direct participants: 2002–2003	31	59.54	52.38	7.16	0.92	6.24

Source: UK ETS transaction log.

target of more than 4,000 companies would not be additional in the ex post sense.

Comparison with Abatement Under a Full Tax Regime

So far, we have assessed the strictness of the CCA targets against the BAU scenario, meaning the scenario with neither CCAs nor the CCL. But if we want to measure the specific contribution of the CCAs to abatement, the relevant benchmark is the level of abatement that would have been obtained under the full tax regime.

As previously noted, ETSU estimated that the CCAs were 10% less effective than the imposition of the full CCL rate. The observation of the ETS market price provides new insights. The market price of the permit lies between £2 and £5, while the level of the tax is equivalent to a price between £5 and £10 per ton of CO_2. It clearly establishes that the CCAs are less effective than the CCL.

From a national interest perspective, this does not necessarily mean that the implementation of CCAs has harmed social welfare because the CCA program also aimed at avoiding possible damage to the United Kingdom's industrial competitiveness.

Cost Effectiveness

Let's complete the assessment by quickly considering abatement cost issues. Key is the connection of the CCAs with the ETS, which provides, in theory, the opportunity to minimize aggregate abatement cost. We have seen that only 20% of the CCA companies have participated in the market due to high transaction (verification) costs. Thus, the ETS has not been able to reap all the gains from trade, and marginal abatement costs probably are not equalized across CCA emissions sources (particularly between buyers and sellers).

Conclusion

The CCAs exhibit an innovative design. They are negotiated agreements embodied in a sophisticated policy mix, as they are combined with a tax exemption and an ETS. The combination with the tax exemption is supposed to secure the environmental strictness of the negotiated targets, while providing a valuable tax-burden reduction to industrial companies subject to harsh international competition. The combination with the ETS should increase the cost-efficiency of the CCAs by providing flexibility to companies.

In four years, the amount of abatement achieved is impressive, as the 2010 target already has been exceeded. However, this huge overachievement may cast some doubt about the genuine environmental strictness of the CCA targets. We develop an original method to assess the environmental additionality exploiting the fact that the ETS price reveals the marginal abatement costs of the CCA participants.

Our conclusion is mitigated. Market prices have remained quite low ($£2–5/tCO_2e$), suggesting low marginal abatement costs. About 1,036 CCA companies have bought permits at this positive price, which proves unambiguously that their target was stricter than the BAU level. They represent around 20% of the CCA companies.

Around 80% of the CCA companies have never traded using the ETS market, while they have exceeded their target by a large amount ($8.8 MtCO_2$). This presumably is due to high verification (transaction) costs, which have hindered their entry into the market. Reduced verification costs would have led to even lower prices. As these overachievements have not been sold on the market, they probably correspond to actions that were profitable and thus not additional.

One may argue that we measure the ex post environmental additionality of the CCAs and not the ex ante additionality the actors had in mind when they were setting the targets. This might be a problem in that ambitious targets in a world of economic fluctuations easily can become slack ex post. The fact that most CCA targets are relative probably mitigates this objection.

Finally, it often is argued that voluntary agreements contribute to the removal of informational or organizational barriers hindering the implementation of profitable abatement actions. If this applies to CCAs, they can yield additional abatement that is not costly ex post. Our methodology fails to address this issue, meaning that we tend to underestimate the CCAs' environmental effectiveness.

References

Atkinson, Tim. 2005. Personal communication between T. Atkinson, consultant at Natsource, and the authors, November 16.

CCX (Chicago Climate Exchange). 2006. http://www.chicagoclimatex.com (accessed August 21, 2006).

DEFRA (Department for Environment, Food and Rural Affairs). 2003. *Transaction Log April 02 -April 03.* http://www.defra.gov.uk/environment/climatechange/trading/uk/reports.htm#2005 (accessed August 21, 2006).

———. 2004. *Transaction Log 01-04-03 to 31-03-04.* http://www.defra.gov.uk/environment/climatechange/trading/uk/ reports.htm#2005 (accessed August 21, 2006).

———. 2005. *Transaction Log 01-04-04 to 31-03-05.* http://www.defra.gov.uk/environment/climatechange/trading/uk/reports.htm#2005 (accessed August 21, 2006).

de Muizon, Gildas, and Glachant Matthieu. 2004. The UK Climate Change Levy Agreements: Combining Negotiated Agreements with Tax and Emission Trading. In *Voluntary Approaches to Climate Policy: An Assessment,* edited by A. Baranzini and P. Thalmann. Ashgate-London: Edward Elgar Publishers, 231–248.

DTI (Department of Trade and Industry). 2004. Updated Emissions Projection: Final Projections to Inform the National Allocation Plan. London: Department of Trade and Industry. http://www.dti.gov.uk/files/file26377.pdf#search=%22Updated% 20emissions%20projection%20%E2%80%93%20final%20projections%20to%20 inform%20the%20National%20Allocation%20Plan%22 (accessed August 21, 2006).

ETSU. 2001. *Climate Change Agreements: Sectoral Energy Efficiency Targets, Version 2.* 01023/3. Harwell, UK: AEA Technology.

Future Energy Solutions AEA Technology. 2004. *Climate Change Agreements: Results of the First Target Period Assessment.* 01838/1. Harwell, UK: AEA Technology.

————. 2005. Climate Change Agreements: Results of the Second Target Period Assessment. AEAT/ENV/R/2025. Harwell, UK: AEA Technology.

Marshall, Lord. 1998. *Economic Instruments and the Business Use of Energy,* Report for the Chancellor of the Exchequer. London: Chancellor of the Exchequer.

OECD (Organisation for Economic Co-operation and Development). 2004. *Voluntary Approaches for Environmental Policy: Effectiveness, Efficiency and Usage in Policy Mix.* Paris: OECD.

Parliamentary Office of Science and Technology. 2004. *Climate Change and Business,* Postnote number 213, January. London, UK: Parliamentary Office of Science and Technology.

Pointcarbon. 2006. www.pointcarbon.com (accessed August 21, 2006).

Segalen, L. 2005. *Etat actuel des marchés CO₂.* Paris: European Carbon Fund and CDC Ixis. www.co2symposium.com/IFP/en/CO2site/presentations/ColloqueCO2_ Session4_05_Segalen_EuropeanCarbonFund.pdf (accessed August 21, 2006).

Notes

1. CERNA, Ecole des Mines de Paris, address for correspondence: 60, boulevard St Michel, 75272 PARIS cedex 06, glachant@ensmp.fr. We thank Ita McMahon and James D. Godber from DEFRA and Tim Atkinson from Natsource for providing useful information. We also thank Dick Morgenstern and William Pizer for commenting on an earlier version of this chapter. The financial support of Resources for the Future is acknowledged.

2. The rate for LPG is even less (£3 per ton of CO_2), but it is a marginal energy source in industry.

3. The Association for the Conservation of Energy was formed in 1981 by a number of major companies active within the energy conservation industry. It aims to encourage a positive national awareness of the need for and benefits of energy conservation. www.ukace.org.

4. As part of the AEA Technology Environment business, Future Energy Solutions (FES) has evolved from the Energy Technology Support Unit (ETSU) set up by the UK Government in 1974. Today it is a private-sector business.

5

Evaluation of the Danish Agreements on Industrial Energy Efficiency

Signe Krarup and Katrin Millock

The overall purpose of the Danish agreements on energy efficiency is to reduce the carbon dioxide (CO_2) emissions from industry to achieve the national objective on CO_2 reduction. The agreements were introduced in 1996 as part of the Green Tax Package that introduced green taxes on industry, trade, and services and a recycling of the tax revenues through subsidies to industry. Tax rebates were given to energy-intensive firms on the condition that they entered into a binding agreement on energy efficiency with the Danish Energy Agency. The agreements do not specify a quantitative target but specify various activities (energy-saving projects, special investigations, and implementation of an energy-management system) that the firm must undertake to qualify for the tax rebate. The agreements are valid for up to three years and can be renewed every third year. The tax rebate has been a strong incentive for energy-intensive firms to conclude an agreement, which has implied a high participation rate by those firms.

The content of the system has been changed several times since the introduction of the agreements in 1996. One reason for changing the system was to lower the relatively high administrative costs related to the first generation of agreements. As a result, the administrative costs of the agreements concluded after 2000 have been reduced. There exist few quantitative evaluations of the environmental effect of the agreements, so the estimation of the effects mainly is based on qualitative evaluation. A government-commissioned evaluation undertaken in 1998 predicted a 6.3% reduction in CO_2 emissions from firms that signed an agreement in 1996–2000. Another evaluation from 2005 estimated the obtained energy savings in firms that have signed an agreement at 2.6% in the 1996–1999 period and 1.9% in the 2000–2003 period. As a comparison, an outside analysis of firms that signed an agreement in 1996–1997 estimated their energy savings to be 4–8%. These evaluations indicate that the agreements have had an effect on firms' behavior, resulting in energy savings and a reduction in CO_2 emissions in industry. The effect has been the largest for the first generation of agreements, and the total effect of the agreements is expected to decrease in the following years. The question then is whether the energy savings would have been achieved without the agreements in place. The evaluations undertaken so far suggest that about half the energy savings obtained in firms with an agreement is due to the agreements.

Background

Following the increased policy attention given to the problem of global climate change (notably the creation of the United Nations Framework Convention on Climate Change in 1992) and the abandonment of the proposal for a European carbon tax, Denmark unilaterally introduced a carbon tax in 1993 to attain its CO_2 emissions reduction target. In its National Energy Plan of 1990, Denmark had stated the objective of a 20% reduction in CO_2 emissions between 1988 and 2005. Further, as an Annex I party to the Kyoto Protocol and according to the EU burden-sharing agreement, Denmark has to reduce its 1990 CO_2 emissions level by 21% for the first commitment period of 2008–2012. The carbon tax was imposed on energy use in trade and industry in 1993 and varied according to the fuel's carbon content. The full rate was EUR 13.3/ton CO_2.[1] Large firms with energy-intensive production could submit themselves to an energy audit and realize certain standard projects. In return, they had their tax payment reduced to a maximum of EUR 1,333 a year regardless of their energy use (and calculated tax) in absolute terms (Johannsen and Larsen 2000). This implies that the effective level of the CO_2 tax on trade and industry was only 35% of the level of taxation on households. Since Denmark is a small, open economy, there were significant concerns about the international competitiveness of Danish industry and potential job losses following heavier energy taxation. A small number of energy-intensive companies account for a large part of all industrial energy use. In fact, approximately 15 companies consume 40% of all industrial energy (Danish Energy Authority 1999). A ministerial committee was set up in 1993 by the government, led by the Social Democrats, to study the possibility of extending green taxes to trade and industry while minimizing any negative impact on international competitiveness. From the very start, all parties agreed on the need to return the revenues from increased energy taxes to industry through reductions in industry's labor market contributions in order to minimize the effects on international competitiveness.

In a preliminary report, the committee suggested that the exemptions for industry should be repealed and that the tax rate on CO_2 for trade and indus-

TABLE 5-1. CO_2 Taxes in EUR Per Ton CO_2

Energy use[a]	1996	1997	1998	1999	2000	2001	2002	2003	2004	2005
Heavy process–no agreement	0.7	1.3	2.0	2.7	3.3	3.3	3.3	3.3	3.3	3.3
Heavy process–with agreement	0.4	0.4	0.4	0.4	0.4	0.4	0.4	0.4	0.4	0.4
Light process–no agreement	6.7	8.0	9.3	10.7	12.0	12.0	12.0	12.0	12.0	12.0
Light process–with agreement	6.7	6.7	6.7	7.7	9.1	9.1	9.1	9.1	9.1	9.1
Space heating[b]		13.3	13.3	13.3	13.3	13.3	13.3	13.3	13.3	13.3

a. For a figure of the full taxes, SO_2 taxes should be added to all energy use.
b. The energy tax should be added to space heating.
Note: Exchange rate: 100 EUR = 750 DKK.
Source: Danish Energy Authority (1999 and 2002a).

try increase from EUR 13.3 to 26.7/ton CO_2 in order to meet Denmark's national and international commitments. Industry protested against the proposal and demanded voluntary measures modeled on the experience of the Dutch Long Term Agreements on Energy Efficiency. The ministerial committee delivered its final report in February 1995 and parliamentary negotiations intensified. Industry's opposition to the increased taxation on energy was supported by many political parties that feared job losses following increased energy taxation on industry. For negotiation purposes in order to have the new energy policy package passed in Parliament, offering an agreement that would give industry access to a lower tax rate was an attractive idea. Agreements had not been used previously in Danish energy policy, and a group of government officials traveled to the Netherlands to study the Dutch agreements. The issue also was complicated by the need to obey European Union (EU) legislation against distortionary measures favoring national industry. Following political discussions among government and industry stakeholders and environmental and consumer organizations, a compromise was reached that resulted in the policy package that came into effect in 1996.

The resulting policy package (called the Green Tax Package) contained a mix of carbon taxes, subsidies for energy-efficiency improvements, and agreements on improved energy efficiency in industry. It aimed at limiting the impact on competitiveness by: 1) increasing tax rates gradually; 2) recycling all revenues directly to industry through a reduction in non-wage labor costs and also through subsidies to energy-efficiency improvements during a transition period; and 3) applying differential tax rates that lower the burden on energy-intensive industries that are subject to foreign competition. The revenues from the energy taxation were recycled to industry by means of reductions in the employer social insurance and pension contributions.

Energy use is liable for three different taxes: CO_2 taxes, sulphur dioxide (SO_2) taxes, and energy taxes. The SO_2 tax was introduced in 1996 and phased in until 2000. It is applied to all energy uses at the same rate, EUR 1.3/kg sulphur (EUR 1,333/ton sulphur), and levied on all fuels with a sulphur content of at least 0.05%. There are no rebates for this tax and signing an agreement on

TABLE 5-2. Estimated Effects of the Green Tax Package by 2005
Compared to 1988 Level

Policy instrument	Reduction (% of total Danish CO_2 emissions)	Reduction (million tons of CO_2 per year)
Tax on space heating	0.8	0.5
Tax on heavy and light processes	0.8	0.5
Agreements	0.6	0.4
Investment grants	1.1	0.7
Tax on SO_2 emissions	1.0	0.6
Total Reduction	4.4	2.8

Source: Johannsen and Larsen (2000).

energy efficiency does not alter the liability for the SO_2 tax. The energy tax is based on energy content and thus varies according to fuel. As part of the new policy package, it was extended to industry and introduced gradually between 1996 and 1998. It has since increased from EUR 5.5/GJ in 1998 to EUR 6.8/GJ in 2002. The level of the Danish CO_2 taxes over time are shown in Table 5-1.

The Specifics of the Danish Agreements on Industrial Energy Efficiency

In 1993, a general CO_2 tax was introduced. However, the most energy-intensive firms were given the opportunity to get their total CO_2 tax payment reimbursed (over an initial sum of EUR 1,333) if they submitted themselves to an energy audit, implemented certain standard projects, and reported their energy-accounting activities to the Danish Energy Agency. This system is sometimes called an exemption system or an agreement system. However, in the following discussion we will not consider this as an agreement.

In 1995, the Danish Parliament adopted a Green Tax Package to reduce total CO_2 emissions by 4.4% of 1988 levels by 2005 (see Table 5-2). The agreements constituted one part of a mixed policy package that also consisted of CO_2 and SO_2 taxes, as well as energy-efficiency subsidies. The expected effect of the agreements was a 0.6% reduction in total CO_2 emissions by 2005 compared to the 1988 level. Two types of agreements were designed from the start: individual agreements and collective agreements, which covered several firms in the same sub-sector. The collective agreements were created to reduce the administrative costs of entering into an agreement and covered firms in the same sector with similar production methods. The agreements were revised in 2000 and 2003; we start by describing the initial features of the policy.

The Danish Agreements 1996–1999

From 1996 to 1999, only energy-intensive firms could enter into an agreement. The Danish Energy Authority separated heavy and light processes. Heavy processes were identified on a process list that defines 35 such processes; for example, production of cement, paper, glass, foodstuff, and greenhouse heating. Light processes were defined as all other energy-using processes other than space heating; for example, energy used for lighting and office machinery. Heavy processes represent 61% of total energy use in Danish industry and light processes account for almost 27%, with the remaining 11% being space heating (Danish Energy Authority 1999). While all firms with heavy processes were defined automatically as energy-intensive, firms with light processes only had the right to enter into an agreement if their yearly tax on energy use amounted to at least 3% of their value added (Danish Energy Authority 1999). This condition effectively excluded many small firms from entering an agreement.

The agreements were qualitative rather than quantitative in the sense that no specific emissions reduction was spelled out. The individual agreements demanded that a firm undertake an energy audit and agree to implement all measures found, with a payback criteria exceeding a given threshold level. In return, the firm received a yearly rebate on its CO_2 tax payment over an initial sum of EUR 1,333 in 1997, EUR 2,000 in 1998–1999, and EUR 2,667 since 2000. Firms covered by a collective agreement received a yearly rebate on their CO_2 tax payment over an initial sum of EUR 1,333 per year. Table 5-1 shows the effective reductions in the CO_2 tax rate for firms with an agreement. The energy audit usually was carried out by independent auditors certified by the Danish Energy Authority, which administers the agreements. The firm paid for the cost of the energy audit but could apply for a subsidy of up to 50% of the actual cost from the Danish Energy Authority. In 1997, the Danish Energy Authority made verification compulsory for all companies. Verification of the energy audit is carried out by agencies that are accredited by the Danish Energy Authority. After verification of the energy audit, the firm outlined an action program based on the energy audit and covering:

1. Energy-saving projects. Firms that entered an agreement for heavy processes had to implement any energy-saving projects with a payback period of less than four years. Firms with light processes had to implement any energy-saving projects with a payback period of less than six years. Energy prices, including taxes at the non-agreement rate, were used when calculating the payback periods. In the calculations, a tax of EUR 3.33 per ton CO_2 was applied to heavy processes, whereas a tax of EUR 12 per ton CO_2 was applied to light processes.

2. Special investigations. If there were energy savings in more complicated processes that could not be thoroughly annualized within the timeframe

established for concluding the agreement, a firm with heavy processes had to implement this project within a year of submission of the result if the payback period was less than four years, while a firm with light processes had to implement the project if the payback period was less than six years.

3. Energy-management measures. The guidelines were formulated in line with other energy-management systems, such as EMAS and ISO 14.001. They outlined how firms could ensure that daily savings were maintained and that opportunities for future savings were evaluated as an integral aspect of daily operations.

Following the energy audit and its verification, the firm negotiated its action program with the Danish Energy Authority. In this phase, alternative projects could substitute for the projects outlined in the initial action program based on the energy audit. Since the Danish Energy Authority does not have other information than what has been provided in the energy audit, the firms have the potential to influence the selection of projects, in particular since the outside consultant that performs the audit relies to a large extent on firm cooperation and information (Krarup, Togeby, and Johannsen 1997). The final agreement listed the firm's obligations in terms of investment projects, special investigations, and implementation of energy-management systems. If no energy-saving projects were identified in the energy audit, the company was considered energy efficient and need not carry out investment projects.

The difference between the individual and collective agreements was that the latter were not based on energy audits of individual firms but on an analysis of energy use and production processes in each sector. Collective agreements have been signed with the following sectors: greenhouses, milk condensation, and structural clay products. In addition to energy saving projects, special investigations, and energy management, the action program could include inter-firm projects, such as research and development (R&D), which were of interest to all firms in the sector. Even though negotiations took place between the Danish Energy Authority and the relevant industrial association, each firm had to sign the agreement separately and had to commit to individual obligations in the action plan. The industrial associations coordinated the annual self-reporting of their member firms to the Danish Energy Authority and were involved in the sector-specific investigations.

Participation in an agreement did not automatically qualify a firm for energy-efficiency subsidies, which were a separate instrument. A company without an agreement could obtain subsidies for energy-saving projects with a payback period of a minimum of two years, whereas a company with an agreement could apply for subsidies for energy-saving projects with a payback period of a minimum of three years. The agreements lasted for three years, and the firm had to supply a yearly report to the Danish Energy Authority. The Danish Energy Authority then followed up on the reports, including renegotiating in case of changed conditions and sanctioning of firms that failed to meet the obligations in the agreement. If a firm failed to meet its obligations, the Danish Energy Authority could cancel the agreement and demand the repayment of the earlier tax reduction. This has occurred twice (Danish Energy Authority 2005). One of the firms did not undertake a special investigation, which was part of their agreement, and the other did not meet the deadline for the conclusion of the agreement.

The Danish Agreements, 2000–2002

In a bill passed in Parliament in December 1999 that came into force in January 2000, the agreements were reformed to reduce the high administrative costs of the instrument and to extend their coverage. The revisions of the agreement scheme concerned the target group and the content and obligations in the agreements. The new generation of Danish agreements emphasizes the adoption of energy-management systems. The energy audit is no longer compulsory and firms must themselves carry out a mapping of their energy use and energy-saving potential to construct an advanced energy-management system. The purpose of the mapping also is to identify areas where special investigations are necessary to identify and implement energy-saving projects with a payback criterion under four years. The target group has been enlarged to firms with a high energy use for space heating. The main part of the monitoring has been delegated to verification agents and technical experts, and the Danish Energy Authority no longer follows up on the firms' self-reports.

The Danish Agreements, 2003 and Onward

The agreement scheme was changed slightly in 2003. First, the target group was once more enlarged to cover firms with a high energy use for hot water. Second, more emphasis was put on the energy-management system, which now has to fulfill specific requirements. When the energy-management system is established, it has to be certified by DANAK, the Danish Accreditation Scheme. Instead of carrying out special investigations, firms now have the possibility to carry out projects on R&D, productivity, and benchmarking. Last, the basic allowance was removed for firms covered by a collective agreement.

The EU member countries have devised a scheme for greenhouse gas emissions allowance trading, effective from 2005. Denmark has ratified the EU directive on greenhouse gas emissions allowance trading and made it law. The energy use covered by the quota system concerns the energy use from heavy and light processes, which implies that the firms that are part of the CO_2 allocation plan are the most energy-intensive ones. These firms get fully reimbursed for their CO_2 tax payment for these types of energy use and have therefore dropped out of their voluntary agreement. Still, these companies retain the possibility of having a voluntary agreement for their energy use for space heating and electricity. By 2005, approximately 60 of the firms with an agreement are part of the Danish allocation plan for CO_2 quotas.

Program Accomplishments: The Views of the Participants

The Sponsoring Agency: The Danish Energy Authority

The expected extraordinary costs for the governmental agencies that administer the agreements amounted to EUR 4 million per year (Rigsrevisionen 1998). In the latest evaluation of the agreements, it was estimated that before 2000, the Danish Energy Authority used 15 man-years to administer the agreements, whereas only half of these man-years were used in the administration of the agreements after 2000 (Petersen, Lorenzen, and Kragerup 2005).

The Danish Energy Authority relies to a large extent on information from the independent consultants that performed the energy audits. Since the introduction of compulsory verification, the quality of the energy audits has increased, according to the Danish Energy Authority. The authority considers its role to be one of accompanying companies toward the adoption of better management systems, and given the high costs of the energy audit, this is the core of the new generation of agreements that came into effect in 2000.

The Danish Energy Authority has commissioned evaluations of the agreements at three occasions (see Krarup, Togeby, and Johannsen 1997; Pedersen et al. 1998; Petersen, Lorenzen, and Kragerup 2005; and Table 5-3). The qualitative study by Krarup, Togeby, and Johannsen (1997) evaluated the experiences with the agreements concluded in 1996. It was estimated that the energy savings agreed upon were 4.8% plus the energy savings caused by the implementation of the energy-management system. A large part of these savings were at that time expected to be achieved even without the agreements in place. As part of the evaluation of the CO_2 tax system, the Danish Energy Agency initiated in 1998 an independent evaluation of the agreement system that was carried out by research institutes and consultants; for example, Pedersen et al. (1998).[2] As the agreement system was not fully established in 1998, it was not possible to undertake an ex post quantitative assessment of the effects of the agreements. Instead the evaluation made use of case studies and a survey of a selected number of firms. The case study of five firms with an agreement focused on how the different elements of the agreements were perceived by the firms. The survey was designed to identify the independent impact of the agreements. It was carried out on 150 firms, of which 91 had signed an agreement in the period 1996–2000. The evaluation predicted that a 6.3% reduction in CO_2 emissions was feasible by 2005 in firms with an agreement compared to a situation without the agreements in place. This corresponds to a total reduction of 3% from the entire industry and to a reduction in total

TABLE 5-3. Assessments of the Expected Impact of the Agreements

Year agreement entered into	Pedersen et al. (1998)	Bjørner and Jensen (2002)	Pedersen, Lorenzen and Kragerup (2005)
1996–1997	2.7% reduction in CO$_2$ emissions	4-8% reduction in energy use	-
1996–1999		–	2.6% reduction in energy use
1996–2000	6.3% reduction in CO$_2$ emissions		
2000–2003	–	–	1.9% reduction in energy use

Danish CO$_2$ emissions of 0.6% of the 1996 level by 2005 (Ministry of Finance 1999). Of the total reduction, each component of the agreements (specific projects, special investigations, energy management) accounted for approximately one third of the reduction during the first three years of the agreements. In later years, energy management systems were expected to provide the largest savings.

The purpose of the latest evaluation of the agreements (Petersen, Lorenzen, and Kragerup 2005) was to evaluate: 1) the total effect of the agreements on the firms' energy use, CO$_2$ emissions, and energy-efficiency performance in the period 1998–2003; 2) how firms experience the administrative costs of the agreements; and 3) to what extent the administrative costs of firms and of the Danish Energy Authority can be reduced. However, due to problems with data (collected by the Danish Energy Authority), it was not possible to make a quantitative assessment of the effects of the agreements. Instead, the effects of the agreements were estimated through a qualitative assessment, including interviews with 28 firms. The evaluation shows that about half of the energy savings obtained since 1996 in these firms have been due to the agreements. This corresponds to net energy savings of 2.6% in the period 1996–1999 and 1.9% in the period 2000–2003. The decreasing effect of the agreements is explained by the fact that the most profitable energy savings were realized with the agreements concluded before 2000. This tendency will continue in the following years, especially when the larger firms opt out of the agreement system due to their participation in the CO$_2$ allowance trading scheme. On this basis, it is therefore expected that the total effect of the agreements will decrease in the future.

Industry: Participation Rates, Cost-effectiveness, and Perceptions

The Confederation of Danish Industry participated actively in the negotiations preceding the introduction of the CO$_2$ tax in 1993. It was mainly industry's opposition to the tax and suggestion of voluntary agreements, based on the Dutch model, which led to the creation of the policy package introduced in 1996. It also has been active in helping companies define their processes as either heavy or light, since this has a significant effect on energy tax liability. As for industry participation, the number of firms with an active agreement in the period 1996–2005 can be seen in Table 5-4. From the table, it can be seen that the number of firms with an agreement increased from 69 in 1996 to 326 firms in 2001, which represents more than 50% of the total energy use in industry (Danish Energy Authority 2002b). After 2001, the number of firms with an agreement has been approximately the same. The firms' participation in the agreements mainly is motivated by the reimbursement of a large part of their CO_2 tax payment. In the latest evaluation, the tax rebates were estimated for most firms (except for greenhouses). The number of firms that have obtained different levels of tax rebates can be seen in Table 5-5.

The administrative costs for the firms have been estimated at EUR 17,000–33,000 on average for each firm. These costs represent the expenses for the energy audit and verification. The Danish Energy Authority (2002a) assessed the administrative costs for the firms at an average of 5–12% of the expected yearly rebate on the CO_2 tax when it is in full effect. In addition, the costs of measuring and defining processes have been estimated to correspond to 1–2% of the tax rebate. See Table 5-6 for estimated administrative costs of the instruments in the policy package. In particular with regard to other instruments, a firm's administrative costs of entering into an agreement have been considered quite high and the agreements were reformed in 2000 to reduce such costs for the firms by canceling the energy audit obligation. In the evaluation made by Petersen, Lorenzen, and Kragerup (2005), the administrative burden to firms is estimated to be 1–3 work months a year and 4–6 work months a year for larger firms. Firms also carry costs related to the implementation of energy-saving projects, certification, and the use of external consultants. The expenses of certification are estimated at EUR 1,333–10,666 per agreement. In the latest evaluation (Petersen, Lorenzen, and Kragerup 2005) it is stressed that smaller firms that receive smaller tax rebates have high administrative costs, amounting to 80–100% of their tax rebates.

Togeby and Hansen (1998) did phone interviews with energy managers from 150 large companies covered by agreements when they were one year into the agreements.[3] The companies thought that the energy audits summed up existing knowledge rather than helping to identify new opportunities to save energy. Energy managers nevertheless felt the agreements had had a positive impact in enhancing their role in the company. In general, the Danish Energy Authority was perceived as flexible with regard to negotiations. Expected energy savings were 2.2%, including investments made without the agreements. There were large differences in expected energy savings between firms, and based on the interviews, it was estimated that 34% of the energy sav-

TABLE 5-4. Number of Firms Part of an Active Agreement[a], Classified by Industry Sector and Type of Agreement

	1996	1997	1998	1999	2000	2001	2002	2003	2004	2005
Individual agreements:										
Mining and quarrying	3	6	7	5	3	3	3	4	3	3
Food and beverages	5	29	47	43	41	29	29	33	32	27
Wood industries	1	3	3	3	2	2	2	1	1	1
Textile industry	0	0	0	0	0	0	0	1	1	1
Paper industries	5	6	6	6	4	4	4	4	4	4
Mineral oil products	1	1	1	2	2	2	2	2	2	2
Chemical industries	5	8	11	10	14	13	14	17	14	11
Rubber and plastic industries	0	2	2	3	1	1	1	1	1	1
Stone, clay, and glass industries	4	7	10	8	15	12	12	9	9	6
Metal manufacturing	1	1	2	4	4	4	4	7	11	9
Iron and steel goods industries	0	4	4	2	3	3	4	1	0	0
Sub-suppliers	4	7	7	5	7	7	7	10	10	10
Sewage work, renovation companies, maintenance companies, etc.	0	0	1	1	0	0	0	0	0	0
Engineering industry	0	0	0	1	0	0	0	0	0	0
Wholesale trade	0	0	0	1	1	0	0	0	0	0
Telecom equipment	0	0	0	0	0	0	0	1	0	0
Real property	0	0	0	0	0	0	0	2	2	2
Entertainment, culture, and sports	0	0	0	0	0	0	0	0	0	1
Total individual agreements	30	76	101	94	97	80	82	92	90	78

Collective agreements:									
Greenhouses	39	81	99	60	215	215	218	195	192	195
Food and beverages	0	9	9	6	7	12	12	17	10	10
Stone, clay, and glass industries	0	0	21	18	18	19	20	19	19	19
Hotels and restaurants	0	0	0	0	0	0	1	9	16	20
Total collective agreements	39	80	129	84	240	246	251	240	237	244
Total agreements	69	166	230	178	337	326	333	332	327	322

a. An active agreement is an agreement that engenders payment of subsidies to cover the CO_2 tax. This means that firms that intend to sign an agreement are included in these numbers.

Sources: Krarup (2002) and the Danish Energy Authority (2005).

TABLE 5-5. The Distribution of Most Firms (Greenhouses Excepted) in Relation to Tax Rebates, 2001

Tax rebate (in EUR)	Number of firms
0–6,667	16
6,667–13,333	14
13,333–40,000	22
40,000–66,667	10
66,667–133,333	11
133,333–266,667	15
266,667–666,667	10
666,667–	2

Source: Petersen, Lorenzen, and Kragerup (2005).

TABLE 5-6. Firms' Administrative Costs of the Instruments in the Green Tax Package

Average annual change 1996–2005	Average annual costs (million EUR)	Average annual reduction in CO_2 emissions (million tons)	Administrative costs compared to CO_2 reductions (EUR/ton)
Taxes	0.9–5.3	0.8	1–7
Investment subsidies	0.7–2.0	0.5	1–4
Agreements	0.9–2.0	0.2	4–8
Entire package	3.1–9.1	1.5	2–6

Source: Danish Energy Authority (2000).

ings would have been implemented without the agreements (Togeby and Hansen 1998).

In 2002, the Danish Energy Authority made a survey on a selected number of firms' experiences with the agreements (Danish Energy Authority 2002c). The conclusions were made from telephone interviews with the energy manager in 20 firms. The answers showed that the majority of the firms had achieved energy savings (or expected to do so) because of the special investigations and the implementation of the energy-management system as part of their agreements. With respect to the administrative costs from implementing the agreements, only half of the firms thought that their costs had decreased after the change in the system after 2000 (which, among other things, tried to reduce the general administrative costs of the agreements).

Environmental Groups: Environmental Effectiveness, Public Perceptions

Environmental and consumer groups were involved only in the beginning of the negotiations on the Green Tax Package of 1995 and did not participate in the negotiations related to the agreements as such; their participation was limited to the discussion of the taxes. The environmental groups were neither involved in the negotiation nor the monitoring of the agreements (cf. Johannsen and Larsen 2000). As agreements are confidential, the exact content and outcome of each agreement is not known to actors other than the firm itself and the Danish Energy Authority. Information on the performance of the agreements only has been available due to the evaluations, and the performance only is available at an aggregate level.

Program Accomplishments: Outside Analysis

An Intermediate Metric:
The Level of Energy-Efficiency Activities within the Company

In a first, early evaluation, the outcome in terms of reduced energy use was not measured, since the agreements only had been in effect for six months. Energy-efficiency activities were instead used as a metric since the agreements were supposed to initiate such activities with the ultimate aim at reducing energy use and CO_2 emissions. Based on data from the phone interviews by Togeby and Hansen (1998), a single rating of the firm's energy efficiency activities was created based on factor analysis and ranging from a scale from 1 (low) to 5 (high). A simple regression analysis explained the energy-efficiency activities in terms of variables linked to company type (energy intensity, sector, size) and policy instruments (e.g., agreements, environmental regulation) that the company had been exposed to (Togeby et al. 1999).

Four variables related to company characteristics and two variables related to policy were found significant in explaining energy efficiency activities. Firms with high energy intensity and/or large companies had a significantly higher level of energy efficiency activities. Also, firms in the food industry or firms with high priority for R&D had higher energy-efficiency activities. As regards policy, companies with an agreement had significantly higher energy-efficiency activities, even when a correction was made for company characteristics. Companies under other environmental regulations but without an agreement showed a medium level of energy-efficiency activities. No conclusions on the actual outcome (in terms of reduced energy demand) can be drawn from this kind of analysis. It is also not possible to compare the effectiveness of different policy instruments, for example between the agreements and an energy tax without exemptions.

Estimations of the Reduction in Energy Demand and CO_2 Emissions

What ultimately matters for target achievement is not improvements in energy efficiency but actual reductions in energy demand. Using a micro-panel database covering the majority of Danish industrial companies over the period 1983–1997, Bjørner and Jensen (2002) completed the most detailed analysis that is available of the effects of the entire Danish policy package (CO_2 tax, agreements, and subsidies for energy-efficiency improvement).

They estimate a single-equation for energy demand at time t with the independent explanatory variables being value added at time t (in fixed prices), the price of energy at time t (in fixed prices), a dummy variable denoting if the company has signed an agreement at time t or earlier, and a variable denoting the accumulated subsidies paid to the company from 1993 to year t as a percentage of the value added of the company. The econometric model includes an individual intercept (or fixed effect) and time dummy variables to control for exogenous technological change. Of the 3,762 companies included in the data, about 9% (348 companies) had at some point obtained a subsidy, while only 2% (60 companies) had had an agreement. The results obtained on the effect of an agreement must thus be interpreted with caution due to the very limited number of firms with an agreement in the sample.

The negative effect on energy demand from an agreement is obtained even on a reduced sample with only the most energy-intensive companies (to account for the fact that only energy-intensive companies could sign an agreement). In a variant of the model with sector-specific demand parameters, Bjørner and Jensen find a reduction in energy demand of 9% from having an agreement, but it was statistically significant only at a 10% level. On the other hand, firms in an agreement receive a tax rebate, which all other things equal, would tend to increase their energy use. This effect is quantified by calculating the effect on the energy price of the tax rebate, which suggests that companies with an agreement have increased their energy use by 1–5% as a direct result of the reduction in the tax rate. A comparison of the two effects suggests that companies with an agreement would have consumed more energy if they had not been offered an agreement but had paid the full tax. Bjørner and Jensen (2002) conclude that the likely net impact on the 60 firms with agreements is a 4–8% reduction in their energy use.

So-called bottom-up studies have found lower effects of the Danish agreements. Bjørner and Jensen (2002) cite the study by Johannsen and Togeby (1999) that estimates a 3% reduction in energy use from the agreements. Bjørner and Jensen suggest that this difference may be explained by the fact that the previous bottom-up studies only included the direct effects of the investment projects specified by the energy audit but did not include the effect of increased energy management as part of the agreement.

To date, the study by Bjørner and Jensen (2002) is the only one that is based on micro-level data on Danish industrial companies and the only one that estimates an explicit demand function for energy use in industry, which enables a comparison of the effect of different instruments. Due to the small number of firms with an agreement in the sample, the results have to be interpreted with care, and the study does not give an estimate of the entire effects of all agreements. Neither does it enable us to compare the relative efficiency of a carbon tax without exemptions and an agreement.

A qualitative study of the effects of the agreements is done in Johannsen and Larsen (2000), who show that firms within the pulp and paper and the milk condensing sectors have carried out the investments and special investigations specified in their agreements. However, only a few firms complied fully with the requirements for the implementation of the energy-management systems. It appears that organizational practices in firms change much more slowly and less effectively than expected. Even though firms comply with the terms of the agreements, the achieved energy savings only have gone a little beyond the baseline.

Environmental Effectiveness, Economic Efficiency, and Administrative Costs

The cited analyses indicate that the Danish agreements have been environmentally effective, although it is not possible to say whether the impact has been higher or lower compared to that estimated in the ex ante analyses. What about the economic efficiency of the allocation of savings across firms? The agreement scheme forces firms to implement investment projects with a payback period of respectively four or six years (before 2000) and less than four years (after 2000). Usually firms accept a payback period of one to two years for energy-efficiency investments, so the agreement scheme forces firms to apply more relaxed criteria when assessing energy-saving investments (cf. Johannsen 2002). The payback criteria used imply that firms with many profitable investments have to realize relatively large savings, and firms with no profitable projects are not loaded with investment projects and special investigations. This mechanism could contribute to an efficient allocation of energy savings between firms. However, according to Johannsen (2002), the energy audits undertaken before 2000 did not entirely eliminate the problem of asymmetric information, which implies that an efficient allocation of energy savings cannot be guaranteed. This implies that projects are mainly carried out when firms suggest them, and this will only occur as part of the overall strategy of a firm rather than because of the agreements.

A complete assessment of the economic efficiency of the agreements cannot be done since the comparison with a full carbon tax without exemptions has not been possible. Most of the evaluation studies that we have reviewed do not attempt to establish a baseline for measuring the program's impact. And when they do so, the studies do not use a common baseline, which makes it difficult to compare them. Notwithstanding, any analysis that aims to measure social welfare under one instrument rather than another has to take into account the administrative costs related to each instrument. Based on the estimates of administrative costs in Table 5-6, it seems that the Danish agreements had much higher relative administrative costs compared to the taxes and the subsidies that also formed part of the policy package. Figures on administrative costs are of course very uncertain, but one may venture at a conclusion in terms of the cost-effectiveness of the instrument: the Danish agreements (in their initial design from 1996–1999) were more expensive than alternative measures in terms of the cost per ton of CO_2 emissions reduction.

Conclusion

The effectiveness of the Danish agreements has to be interpreted in the light of them being part of a policy package consisting of CO_2, SO_2, and energy taxes, as well as energy-efficiency subsidies. The Danish agreements were designed to complement the CO_2 tax, not as an instrument on its own (as opposed to the Long Term Agreements in the Netherlands). In view of Denmark's commitments to climate change policy, with a national reduction goal of 20% of 1988 emissions by 2005, the introduction of the policy package of 1995 came as no surprise to the actors. The threat of a carbon tax was very real and it is indeed the linkage between the tax and the agreements that has been the main factor in explaining industry's participation in the agreements. The Danish agreement contained a clear and explicit sanction: if a firm failed to comply with its agreement, it had to repay the CO_2 tax rebate. In addition, it is worth noting that firms have to pay SO_2 taxes regardless of any agreement and such taxes affect the profitability of the energy-saving projects identified in the agreements.

A main characteristic of the Danish agreements is that enforcement rests entirely with the Danish Energy Authority. There is no potential for free riding in that agreements and tax reductions are individual benefits following the commitment to an individual action plan. The agreements have been flexible in that they have been reformed over the years following experience and input from participants. Whereas the initial agreements from 1996–1999 were based on the energy-saving projects that were identified in an energy audit, the agreements of the year 2000 and onwards do not comprise a compulsory energy audit but are based on the implementation of energy-management systems.

In total, the agreements have led to energy savings and CO_2 reductions in the Danish industrial sector. One evaluation undertaken in 1998 predicted a 6.3% reduction in CO_2 emissions from firms that signed an agreement in 1996–2000 and another estimated the obtained energy savings in firms that have signed an agreement to be 2.6% in the period 1996–1999 and 1.9% in the period 2000–2003. This suggests that their effect has been decreasing since they were introduced in 1996. The first generation of the agreements made firms implement the most profitable energy-saving potential, whereas further savings only can be obtained with development of new technology. In addition, the most energy-intensive firms will opt out of their voluntary agreements because of the European trading system with CO_2 quotas. This suggests that the future role for agreements in their present form is small and new measures must be found if further energy savings are to be obtained in industry.

References

Bjørner, T.B., and H. H. Jensen. 2002. Energy Taxes, Voluntary Agreements and Investment Subsidies: A Micro Panel Analysis of the Effect on Danish Industrial Companies' Energy Demand. *Resource and Energy Economics* 24(3): 229–249.

Danish Energy Authority. 1999. *The Danish Agreements on Energy Efficiency*. Copenhagen: Danish Energy Authority.

———. 2000. *Green Taxes for Trade and Industry: Description and Evaluation*. Copenhagen: Danish Energy Authority.

———. 2002a. *Green Taxes in Trade and Industry: Danish Experiences*. Copenhagen: Danish Energy Authority.

———. 2002b. *Voluntary Agreements on Energy Efficiency: Danish Experiences*. Copenhagen: Danish Energy Authority.

———. 2002c. *Opfølgning af aftaleordningen: erfaringer fra 20 virksomheder (Assessment of the Agreement Scheme: Experience from 20 Firms)*. Copenhagen: Danish Energy Authority.

———. 2005. Email correspondence.

Johannsen, K.S. 2002. Combining Voluntary Agreements and Taxes: An Evaluation of the Danish Agreement Scheme on Energy Efficiency in Industry. *Journal of Cleaner Production* 10: 129–141.

Johannsen, K.S., and A. Larsen. 2000. *Voluntary Agreements: Implementation and Efficiency, The Danish Country Study. Case Studies in the Sectors of Paper and Milk Condensing*. Copenhagen: AKF Forlaget.

Johannsen, K.S., and M. Togeby. 1999. Evaluations of the Danish CO_2 Agreement Scheme. European Network on Voluntary Approaches (CAVA), CAVA Working Paper 98/11/7.

Krarup, S. 2002. *Voluntary Approaches: Two Danish Cases*. ENV/EPOC/WPNEP (2002) 13/FINAL. Paris: OECD.

Krarup, S., and S. Ramesohl. 2000. *Voluntary Agreements in Energy Policy: Implementation and Efficiency, Final Report from the Project Voluntary Agreements: Implementation and Efficiency (VAIE)*. Copenhagen: AKF Forlaget.

Krarup, S., M. Togeby, and K. Johannsen. 1997. *De første aftaler om energieffektivisering-Erfaringer fra 30 aftaler indgået i 1996 (The First Agreements on Energy Efficiency: Experience from 30 Agreements Concluded in 1996)*. Copenhagen: AKF Forlaget.

Ministry of Finance. 1999. Evaluering af grønne afgifter og erhverveve (Evaluation of the Green Tax System). Copenhagen: Ministry of Finance.

Pedersen, P.B., C. Ingerslev, M. Togeby, and G. Ahé. 1998. *Evaluering af energiaftalernes effekt (Evaluation of the Impact of the Energy Agreements)*, Dansk Energi Analyse. Copenhagen: AKF, Rambøll.

Petersen, P.M., K.H. Lorenzen, and H. Kragerup. 2005. *Evaluering af aftaleordningen om energieffektivisering 1998-2003 (Evaluation of the Agreement Scheme on Energy Efficiency 1998-2003)*. Copenhagen: COWI.

Rigsrevisionen. 1998. *Beretning om indførelse, administration og kontrol af den grønne afgiftspakke (De af Folketinget Valgte Statsrevisorer 2/98) (Report on the Implementation, Administration and Monitoring of the Green Tax Package)*. Copenhagen: Rigsrevisionen.

Togeby, M. and E. Hansen. 1998. Industriens energieffektiviteter—resultater fra 150 virksamheder (Industrial Energy Efficiency—Results from 150 Firms). Copenhagen: AKF.

Togeby, M., T. B. Bjørner, and K. Johannsen. 1998. *Evaluation of the Danish CO_2 Taxes and Agreements*. Copenhagen: AKF.

Togeby, M., K. Johannsen, C. Ingerslev, K. Thingvad, and J. Madsen. 1999. Evaluations of the Danish Agreement System. Paper presented at the ACEEE Summer Study in Energy Efficiency in Industry, Saratoga Springs, NY.

Notes

1. Throughout this chapter, we use an exchange rate of 100 EUR=750 DKK.

2. The evaluation is also summarized in Ministry of Finance (1999).

3. The sample consisted of 31 companies with old agreements (from 1993–1995), 91 companies with new agreements (from 1996–1998), and 28 other companies (without agreements).

6

Assessing Voluntary Commitments in the German Cement Industry

The Importance of Baselines

Christoph Böhringer and Manuel Frondel

Voluntary agreements have become a vital alternative to mandatory policies based on regulation or legislation, specifically in the field of environmental protection. For example, in addition to a variety of voluntary farm programs, several federal programs have been designed in the United States to convince firms to reduce pollution voluntarily (see Wu and Babcock 1999). In the European Union, the majority of Member States rely on environmental agreements as a policy tool, with a wide range of applications, including water and air pollution and waste management (EC 1996). In the Organisation for Economic Co-operation and Development countries, more than 350 voluntary schemes have been established, mainly in Germany and the Netherlands (OECD 2000).

The increasing popularity of voluntary approaches raises two fundamental questions. The first is whether voluntary approaches generally are effective. The study of Segerson and Miceli (1998) is one of the most recent theoretical papers that deal with this first issue of the effectiveness of voluntary approaches in the context of environmental protection. If effectiveness is guaranteed, a second task is to evaluate how efficient voluntary approaches are compared with other approaches. Alberini and Segerson (2002) consider both the environmental effectiveness and the efficiency of voluntary approaches and provide an excellent discussion of theoretical and empirical issues that arise in the evaluation or assessment of a particular voluntary agreement.

Generally, three types of voluntary agreements can be distinguished (see e.g., Alberini and Segerson 2002). The first type induces participation by providing positive incentives, such as cost-sharing (carrot approaches). The importance of these incentives is stressed by Wu and Babcock (1995) and more recently by Lyon and Maxwell (2003). Alternatively, participation can be achieved by threatening a harsh, legislative compulsion (stick approaches). In approaches of this type, agreements on issues such as levels of environmental protection typically are negotiated between a whole industry, represented by an industrial organization, and a regulatory agency, leaving it to the industry itself to solve coordination and free-riding problems.

In other words, a voluntary agreement of this second type requires mutual acceptance of the terms, yet without any legal obligation for the firms of such an industry. This type of agreement often is said not to be truly voluntary, because an industry essentially is choosing the lesser of two evils (see e.g., Goodin 1986). It has been argued that the differences between such agreements and direct regulation are trivial, since the regulator may simply threaten to impose extremely harsh policy measures to enforce the acceptance of the agreement. Hence, some authors prefer the term "negotiated" instead of "voluntary" agreements (Nyborg 2000); for a recent survey of such voluntary agreements, see Khanna (2001).

This chapter focuses on the effectiveness of a third type of voluntary approach which we call voluntary commitment. Its particular characteristic is that it represents a unilateral declaration without a decisively active role on the part of regulators.[1] By definition, voluntary commitments are not the result of intensive mutual negotiations between participants and regulators. Such an approach is defined here to be environmentally ineffective if the actual outcome of the commitment measure, such as specific emissions, mimics business as usual; that is, the counterfactual outcome that would have occurred in the absence of this policy instrument, while all other conditions varied as they actually did. By combining theoretical considerations of the economic rationale for the popularity of voluntary commitments with an investigation of the principal conceptual and statistical problems regarding their empirical assessment, most notably that counterfactual situations principally are unobservable, we cast doubt on the environmental effectiveness of this specific type of voluntary approach.

A prominent example of a voluntary approach in Germany is the "Declaration of German Industry on Global Warming Prevention" (GGWP declaration) of 2000, originally initiated in 1995.[2] Specifically, it is the first version of the GGWP declaration that may be classified as a voluntary commitment, as updates of the original declaration were at least partly due to government intervention. Using the example of the regularly monitored GGWP declaration, we theoretically argue that a fundamental difficulty in assessing the effectiveness of such voluntary approaches is the construction of the appropriate counterfactual scenario describing what would have happened in the absence of these instruments.

Our empirical illustration of this argument draws on data from the cement industry, which provides the only, albeit poor, data source available for this issue. The German cement industry committed itself in March 1996 to reducing its specific fuel consumption by 2005 by 20% relative to the level of the base year 1987. In 2000, the cement industry additionally committed itself to reduce the energy-related specific CO_2 emissions by 2012 by 28% relative to the level of the new base year 1990. In the course of the monitoring of the GGWP declarations, post-announcement data have been gathered by all involved industrial associations to control the progress in reaching environmental targets, whereas longitudinal data of a long range prior to the declarations is scant. The energy data provided by the cement industry represent the sole exception.

Investigating the historical specific fuel consumption of this industry, we find that energy- and emissions-reducing activities have not gone much beyond good intentions. Indeed, despite the monitoring of the declaration, there does not seem to be a significant deviation from business-as-usual. Yet, while we must confine our empirical illustration to the example of the cement industry, we argue that our theoretical reasoning and discussion of conceptual problems with respect to voluntary approaches of the voluntary commitment type holds in general. In short, it is the impossibility of observing the counterfactual, along with the absence of data that could be used to construct a credible counterfactual, that make it relatively easy for the industries involved in the GGWP declaration or other proponents to declare the effort a success.

Below we explain the popularity of voluntary commitments, followed by a discussion of their principal conceptual, statistical, and empirical problems as illustrated by the German cement industry.

The Appeal of Voluntary Commitments

In this section, it will be illuminated from the theoretical perspective of rational economic behavior why voluntary commitments—to reduce CO_2 emissions, for instance—are generally very attractive for both politicians and governments on the one hand, and firms and industrial organizations on the other. For politicians and legislators, there are numerous reasons why voluntary commitments are appealing. First, it is often argued in the economic literature that voluntary approaches generally are vital competitors to mandatory legislation and regulation policies because they guarantee the achievement of environmental quality goals at lower cost. In the absence of any negotiations, monitoring costs appear to be the only administrative and transaction costs of voluntary commitments, while environmental compliance costs are believed to be much lower relative to inflexible regulatory and legislative approaches (see e.g., EC 1996). As opposed to regulatory approaches that dictate inefficient abatement strategies, the greater flexibility of voluntary approaches represents a potential for cost savings.

Besides these theoretical arguments, the most appealing advantage of voluntary commitments from a politician's practical point of view might be that their own popularity may not suffer at all, or at least not as much, as it would with mandatory approaches. The German government, for example, appreciates, according to Alberini and Segerson (2002), the GGWP declaration as an opportunity to fulfill its national CO_2 reduction targets without placing a heavy burden on German industries. Moreover, in cases of non-compliance, politicians easily can defer much of the weight of responsibility to the industries.

With respect to the cost of voluntary approaches, to the best of our knowledge, there does not seem to have been a single empirical or statistical analysis on this issue. This simply might be the result of the inherent principal difficulty in policy assessment that will be discussed in the following paragraphs: the determination of the counterfactual benchmark scenario that is indispensable to both measure the environmental performance and the cost-effectiveness of voluntary approaches (see Alberini and Segerson 2002). It is a key difficulty for the evaluation of the success of any policy instrument, however, that counterfactual situations, precisely describing what would have happened in the absence of this instrument, are inherently unobservable.

Out of the multitude of counterfactual situations, with stopping production and emissions being an unrealistic option in the GGWP example, two alternative counterfactual reference scenarios usually are adopted for the evaluation of a single instrument. First, the fixed-technology scenario, which fixes the status quo prior to the policy instrument. It is very unlikely, though, that this scenario coincides with reality even if economic circumstances remain unchanged. In the example of the GGWP declaration, for instance, autonomous energy-efficiency improvements, which are neither triggered by economic conditions nor by any policy instrument but are the result of a steady march of technological progress in production, usually vary the status quo of the level of energy consumption and emissions.

Typically, however, it is due to altered economic circumstances, definitely not a consequence of the policy instrument alone, that the status quo changes. Yet, when choosing the fixed-technology scenario for the assessment of the effects of a policy instrument, one necessarily has to assume that any changes in the respective outcome measure are the sole result of the policy instrument, not of changing economic conditions. This assumption is hardly true. Hence, this scenario represents an inadequate benchmark for correctly assessing the effects of any policy instrument. Nevertheless, this benchmark often is selected when no empirical data prior to the instrument are available. It is also the scenario that is implicitly assumed in the monitoring of the GGWP declaration, where post-announcement data are gathered, while data prior to the declaration are hardly available (see section three).

The second reference scenario is the business-as-usual scenario, reflecting the development that would occur if, apart from the introduction of the instrument, all other conditions—prices, in particular—varied as they actually did. Although not observable, this counterfactual scenario often is identified by an extrapolation of the historical development, provided that the required data prior to the policy instrument are available. The European Environment Agency (1997) thus calls it the trend scenario, appearing to be unequivocal, which it is not, since extrapolating the historical development by fitting trend curves generally may allow for ambiguity (see section three).

To explain the special appeal of voluntary commitment for firms and industries, we now hypothesize that voluntary commitments represent little more than commitment to business-as-usual, which represents all activities that would also have been pursued in the absence of voluntary commitments. Given the fact that counterfactual situations are inherently unobservable, and that appropriate counterfactual baseline situations—which are indispensable for evaluation purposes—are thus difficult to identify, it is hard for outsiders, such as politicians or scientists, to decide whether these commitments differ from business-as-usual. In other words, the impossibility of observing counterfactual developments and the difficulty of getting a judgement on the most likely alternative development—something that even is hard to appraise for the firms and industries themselves—make it easy for them to call the commitment a real effort.

Provided that our working hypothesis holds true and given the situation of asymmetric information between firms and industrial organizations on the one side and consumers, competitors, and public authorities on the other, voluntary commitments seem to be a sensible reaction of rational economic agents such as firms and industrial organizations. In those situations where politicians discuss the implementation of various mandatory policy interventions, the strategic objective pursued by firms and industries via voluntary commitments is to avoid, or at least delay, costly mandatory policy interventions to secure business-as-usual profits. Exactly these benefits might represent the principal cause for the declaration of voluntary commitments. Moreover, it seems unlikely that voluntary commitments of economic agents, knowing that the goals of the commitments are difficult to distinguish from business-as-usual, induce a significantly more ambitious level of environmental quality than in the counterfactual situation without such commitments.

A variety of ancillary incentives may increase the attraction of voluntary commitments for firms and industries; see Alberini and Segerson (2002) for an exhaustive discussion on incentives for participation in voluntary approaches. First among these and probably most important, voluntary environmental regulation can be considered part of a firm's or industry's public relation activities (see e.g., Arora and Cason 1996). According to Cavaliere (2000), environmental reputation effects can be regarded as implicit contracts and are frequently self-enforcing. Second, struggles going beyond business-as-usual still could be possible. They might originate from unrealized no-regrets options; that is, from those strategies that reduce both production cost and emissions—for example, by energy-conservation measures. Due to negative reputation effects in the case of non-compliance, the public declaration of environmental goals in a voluntary commitment might support a more disciplinary realization of already perceived no-regrets potentials or might even initiate an intensified search for unperceived potentials. Exactly those no-regrets potentials that are merely realized or perceived as a consequence of the commitment appear to represent the potential of the environmental effectiveness of voluntary commitments.

In sum, taking on the perspective entailed in our hypothesis—its validity notwithstanding—may reveal the genuine intentions behind voluntary commitments. For industries that in effect commit themselves to little more than business-as-usual, this kind of policy instrument may provide for net benefits originating from the delay—or even the avoidance—of the actual imposition of mandatory economic threats such as a carbon/energy tax.[3] From a theoretical perspective, our hypothesis does not seem to be unreasonable. Private firms should have no incentive to provide a public good, like climate protection, unless it is in their self-interest, which is the case for no-regrets measures inducing cost-savings and, hence, net benefits, or unless there are credible mandatory economic threats.

The conceptual, statistical, and empirical problems in distinguishing whether or not actual performance differs from business-as-usual and subsequently in verifying the environmental effectiveness of voluntary measures may make voluntary commitments an effective strategic means to weaken the relevance of a potential regulatory pressure. Based on the descriptive trend analysis of the following section, our working hypothesis cannot be rejected in the specific example of the German cement industry.

Empirical Illustration

In 1995, the Federation of German Industries declared the target of the reduction of CO_2 emissions by up to 20% of 1987 levels by the year 2005. Due to the variability of the upper limit, this GGWP declaration was nothing more than symbolic. In 1996, as a consequence of severe criticism of the government, the target was changed to a definite 20% reduction relative to a new base year, 1990, which is in line with international practice. Moreover, the declaration was joined by additional associations. With a total of 19 industry associations involved today, it encompasses more than 70% of final industrial energy consumption and almost the entire public electricity generation of Germany. In 2000, when most of the commitments of the individual industrial associations already were fulfilled within half of the period, the GGWP declaration was again updated, also as a consequence of government influence. The target was revised to a 28% reduction relative to 1990 by the year 2005. Additionally, a new target for specific emissions of other greenhouse gases was recommended: the reduction of the emissions of six specific gases by 35% relative to the 1990 level by the year 2012.

The German cement industry committed itself in March 1996 to reducing its specific fuel consumption by 20% relative to the base year 1987 by the year 2005. In absolute terms, this target implies a specific fuel consumption of 2,800 kJ per kg cement in 2005. In 2000, the cement industry additionally committed itself to reducing the energy-related specific CO_2 emissions by 28% relative to the new base year of 1990 by 2012. In the course of the monitoring of the GGWP declarations, post-announcement data have been gathered to control the progress in reaching environmental targets, whereas longitudinal data of a long range prior to the declarations has hardly been available. The energy data provided by the cement industry represents the sole exception.

Based on the historical energy data, we now evaluate the impact of the energy-related declaration of the cement industry relative to the period before its announcement by a so-called before-after comparison. Due to the lack of sufficient data, unfortunately, more sophisticated methods such as a difference-in-difference approach cannot be applied (for a summary of evaluation approaches in environmental contexts, see Frondel and Schmidt 2005). A before-after comparison, one of several approaches to the solution of the classical evaluation problem, consists of the construction of an appropriate counterfactual situation, which means a precise statement of what would have happened in the absence of a policy intervention. Although not observable, the business-as-usual scenario—the relevant counterfactual situation invoked by our hypothesis—is identified by estimating the historical trend of the cement industry's specific fuel consumption,

$$\log fuel = \log fuel_{1974} - r \cdot trend + \varepsilon, \tag{1}$$

and predicting specific fuel consumption for the near future, that is, up to 2005, by extrapolating the historical-specific consumption curve. The estimation and extrapolation of the historical trend amounts to forecasting by "sighting along the data," as Schmalensee, Stoker, and Judson (1998) call it. In model (1), the variable *fuel* denotes specific fuel consumption of cement production, measured by the fuel input in kJ per kg cement, and *trend* stands for a linear time trend taking on the values 1974, 1975, . . . , 1995. An estimate of parameter r yields the estimated rate of fuel efficiency improvements, when energy efficiency is captured in terms of specific energy input.

The descriptive exponential growth model (1) illustrates a very simple route for an industrial association to define its commitment targets. Deliberately, we abstain in model (1) from additional variables. While climatic conditions, for instance, are important in the textile industry, they should not play any role in the cement industry and, hence, are not incorporated. Business-cycle effects can be expected to influence absolute rather than specific fuel consumption. In addition, business cycles do not seem to vary absolute cement output and, hence, absolute fuel consumption substantially. Thus, proxies of economic conditions such as GDP are not included in model (1) either.

Clearly, structural changes, such as German reunification, cannot be ignored even in simple trend description, nor can major economic changes such as the oil-price shocks in the 1970s. Figure 6-1, reporting time-series data (1960–1999) on the specific fuel consumption of German cement production, displays a drastic decrease of specific fuel consumption in 1974. It roughly amounts to 80% of 1973's specific fuel consumption. Figure 6-1 indicates that there seems to be an attenuation effect, with specific fuel consumption being much lower in the period after the oil price shocks. Because of this attenuation effect and the likely persistence of the oil crises, we ignore data before the first oil price shock in 1973/74.

Overall, for the estimation of model (1), we employ yearly specific fuel data from 1974 until 1995, the year of the first announcement of the GGWP declaration. To capture reunification effects, a reunification dummy should be added to model (1). Yet, this dummy turns out to be insignificant. When thus ignoring this dummy, estimates of the parameters of model (1) are obtained by using Ordinary Least Squares (OLS):

$$\log fuel = 20.642 - 0.0063 \cdot trend \qquad (2)$$
$$(1,952) \quad (0.00098)$$

Standard errors of the OLS estimates are given in parentheses. On average, yearly fuel efficiency improvements amount to 0.63% within the period 1974–1995. By contrast, average yearly fuel efficiency improvements are around 1.3% within the whole period from 1960–1995, where the years before the oil-price shocks are included.

Alternatively, a linear trend model performs equally well in terms of fit and prediction,

$$fuel = 43636.1 - 20.37606 \cdot trend, \qquad (3)$$
$$(6332.9) \quad (3.19118)$$

indicating an average decline of roughly 20 kJ per kg in specific fuel consumption per year, which corresponds to the yearly fuel efficiency improvements of 0.63% for the same period, 1974–1995. The presentation of the alternative model (3) illustrates that the identification of the business-as-usual scenario by estimating historical trends clearly is not unequivocal. For interpretational convenience, for example, one might prefer model (1). In short, although the simplicity of these descriptive models mainly has been dictated by the lack of data, they are plausible in the case of the cement industry.

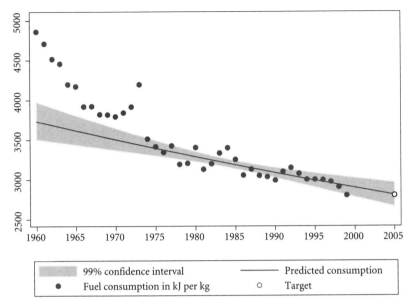

FIGURE 6-1. Actual Specific Fuel Consumption of the German Cement Industry (1960–1999) in kJ per kg and Its Prediction for 1996–2005

Sources: Butterman (1997), Butterman and Hillenbrand (2000).

According to the classical way of statistical testing, we take the hypothesis of section two as the null hypothesis: The actual development of specific fuel consumption after the declaration in 1995 more or less reflects the business-as-usual scenario given by the extrapolation based on model (3). The null is simply tested by using a forecast interval for the 99% confidence level, for example, which is depicted in Figure 6-1. The lower bound of the interval amounts to 2,666 kJ per kg for the target year 2005, the upper bound is 2,956 kJ per kg. Hence, the target of 2,800 kJ per kg is located almost in the center of this interval.[4] Obviously, with the business-as-usual prediction hitting the commitment target almost perfectly, the null cannot be rejected irrespective of the selected significance level. In other words, the commitment target seems to be perfectly consistent with historical experience. This leads to the conclusion that to achieve substantial environmental effectiveness, a more demanding exogenous goal should have been negotiated between government and the cement industry.

In sum, the voluntary commitment of the industrial association of German cement producers hardly seems to be environmentally effective because trends in specific fuel consumption appear to follow business-as-usual. Hence, this example is an illustration of our working hypothesis. The monitoring report of the German cement industry is the sole report that allows a rough evaluation of its commitment. As a result, the genuine environmental effects of the commitments of all other industries involved in the GGWP declaration cannot be discovered without data prior to the declaration because these data are indispensable for evaluation approaches such as before-after comparisons. Post-announcement data provided by the monitoring reports are not enough for the construction of the counterfactual situation. The grave lack of evaluation data prior to the GGWP declaration makes impossible a scientifically based decision on whether the environmental performance of all other industries differs from business-as-usual.

Given that business-as-usual is what is intended, however, it would clearly not be the highest priority of industrial associations to alleviate information deficits with respect to pre-announcement data. Yet, to convince both the German government and the public that voluntary commitments are environmentally effective, more information is required than heretofore documented in the monitoring reports. If pre-announcement information is not available at the individual industry-level, which seems unlikely, the question is how individual commitment targets could have been determined.

Summary and Conclusion

The term "voluntary commitment" is chosen here for a specific type of unilateral voluntary approach having the particular characteristic that it is not the result of decisive government interventions; that is, mutual negotiations between participants and regulators. This particular voluntary approach lies at the heart of this chapter. We hypothesize that unilateral voluntary declarations might represent the commitment to, in effect, more or less maintain business-as-usual and argue that it would be extremely difficult for politicians or any other outsider to decide whether actual environmental performance differs from business-as-usual.

Notwithstanding its empirical validity, the perspective entailed in this hypothesis reveals that a voluntary commitment can be an effective strategy by industrial associations to delay or even circumvent regulatory threats. It is much more likely that declaration targets will go beyond business-as-usual if regulators play an active role; that is, if the goals to be achieved are the result of intensive mutual negotiation between regulators and participants. Mutual negotiation might help to make goals more demanding so that potential regulatory threats appear to be more clear, strong, and credible. These are key features that, according to Alberini and Segerson (2002), are likely to increase the effectiveness and efficiency of voluntary approaches.

A prominent example of voluntary approaches is the first phase of the GGWP declaration. While it does not represent a proper example of a voluntary commitment in its pure form, it shows that regular monitoring must not be confused with a thorough assessment of the voluntary approach. Our argument that a judgement on the environmental effectiveness of voluntary commitments is difficult for outsiders has been supported here by the discussion of conceptual, statistical, and empirical problems involved with the assessment of voluntary approaches. Above all, the lack of pre-declaration data is a key obstacle, highlighting that post-announcement monitoring like the one in the GGWP example is not sufficient to guarantee environmental effectiveness. Rather, negotiations—with carrots and/or sticks—are necessary. Thus, from the vantage point of both our theoretical and empirical arguments, it seems unlikely that voluntary approaches of the voluntary-commitment type are able to induce significant deviations from business-as-usual. This casts doubt on the effectiveness and, hence, the efficiency, of voluntary commitments.

Our arguments are illustrated by an empirical investigation of the declaration of the German cement industry, the only industry where pre-declaration data are available. In line with conclusions of previous studies on the overall voluntary commitment of German industries (see e.g., EEA 1997), we find little evidence that the commitment of the German cement industry is effective—that is, leads to the reduction of specific energy consumption significantly below the business-as-usual level. Voluntary commitments at the industry rather than the firm level, as in the GGWP declaration, are further weakened by diminished incentives to improve or threaten firms' public image. In conclusion, therefore, voluntary approaches should involve negotiated and firm-specific commitment targets to be more effective and should require both a pre- and post-announcement monitoring.

References

Alberini, A., and K. Segerson. 2002. Assessing Voluntary Programs to Improve Environmental Quality. *Environmental and Resource Economics* 22: 157–184.

Arora, S., and T.N. Cason. 1996. Why Do Firms Volunteer to Exceed Environmental Regulations? Understanding Participation in EPA's 33/50 Program. *Land Economics* 72: 413–432.

Böhringer, C. 2002. Climate Politics From Kyoto to Bonn: From Little to Nothing. *The Energy Journal* 23(2): 51–71.

Buttermann, H.G. 1997. Ein Modell zur Erklärung des Faktoreinsatzes in der deutschen Zementindustrie. RWI-Papiere 48. Essen: RWI.

Buttermann, H.G., and B. Hillebrand. 2000. Third Monitoring Report: CO_2 Emissions in German Industry 1997–1998. RWI-Papiere 70. Essen: RWI.

Cavaliere, A. 2000. Overcompliance and Voluntary Agreements. *Environmental and Resource Economics* 17: 195–202.

EC (Commission of the European Communities). 1996. *On Environmental Agreements.* Communication from the Commission to the Council and the European Parliament. Brussels: EC.

EEA (European Environment Agency). 1997. Environmental Agreements: Environmental Effectiveness. *Environmental Issues Series* 1(3). Copenhagen: EEA.

Frondel, M., and C.M. Schmidt. 2005. Evaluating Environmental Programs: The Perspective of Modern Evaluation Research. *Ecological Economics* 55(4): 515–526.

Goodin, R.E. 1986. The Principle of Voluntary Agreement. *Public Administration* 64: 435–444.

Jochem, E., and W. Eichhammer. 1999. Voluntary Agreements as an Instrument to Substitute Regulations and Economic Instruments: Lessons from the German Voluntary Agreements on CO_2 Reduction. In *Voluntary Approaches in Environmental Policy*, edited by C. Carraro and F. Leveque. Dordrecht: Kluwer Academic Publishers, 209–227.

Khanna, M. 2001. Non-Mandatory Approaches to Environmental Protection. *Journal of Economic Surveys* 15(3): 291–324.

Lyon, T.P., and J.W. Maxwell. 2003. Self-Regulation, Taxation and Public Voluntary Environmental Agreements. *Journal of Public Economics* 87: 1453–1486.

Nyborg, K. 2000. Voluntary Agreements and Non-Verifiable Emissions. *Environmental and Resource Economics* 17: 125–144.

OECD (Organisation for Economic Co-operation and Development). 2000. *Voluntary Approaches to Environmental Policy: An Assessment*. Paris: OECD.

Schmalensee, R., T.M. Stoker, and R.A. Judson. 1998. World Carbon Dioxide Emissions: 1950-2050. *Review of Economics and Statistics* 80: 15–27.

Segerson, K., and T.J. Miceli. 1998. Voluntary Environmental Agreements: Good or Bad News for Environmental Protection? *Journal of Environmental Economics and Management* 36: 109–130.

Wu, J., and B.A. Babcock. 1995. Optimal Design of a Voluntary "Green Payment" Program Under Asymmetric Information. *Journal of Agricultural and Resource Economics* 20(2): 316–327.

———. 1999. The Relative Efficiency of Voluntary vs. Mandatory Environmental Regulations. *Journal of Environmental Economics and Management* 38(2): 158–175.

Notes

1. Besides the carrot approach, the unilateral declaration is a widespread type of voluntary approach in the United States; see OECD (2000).

2. German industry committed itself to voluntarily reducing carbon dioxide (CO_2) emissions and/or energy consumption. These activities, often termed "business-led initiatives," "corporate environmentalism," or "industry self-regulation" (see Alberini and Segerson 2002) represent a voluntary unilateral contribution to anticipate the abatement efforts demanded by the Kyoto Protocol. In its original version, industrialized countries are supposed to reduce greenhouse gas emissions during the 2008–2012 period by an average of 5.2% below their 1990 level (see e.g., Böhringer 2002).

3. In the GGWP example, a heat recovery ordinance was the mandatory threat that the German government contemplated to meet the emissions targets fixed in the Kyoto Protocol (see Jochem and Eichhammer 1999). Regardless of the concrete mandatory alternative that might be implemented, however, firms and industrial organizations, particularly of energy-intensive industries, are expecting a priori substantially higher production costs.

4. Note that these bounds are based on the assumption that the development of the specific fuel consumption follows a stationary process. If this does not hold true, the bounds of the interval are wider, but the basic conclusion that the commitment is business-as-usual remains unchanged.

7

Evaluating Voluntary U.S. Climate Programs
The Case of Climate Wise

Richard D. Morgenstern, William A. Pizer,
and Jhih-Shyang Shih[1]

Voluntary programs have been a key part of U.S. climate change policy for more than a decade. Such programs figured prominently in President George W. Bush's 2002 climate change policy announcement, referencing recent agreements with the semi-conductor and aluminum industries and leading to the creation of the Climate Leaders and Climate Vision programs (White House 2002, 2005). They also were the centerpiece of President Clinton's 1993 Climate Change Action Plan, which included Energy Star, Rebuild America, Green Lights, Motor Challenge—and Climate Wise (Clinton and Gore 1993). Further, as this volume points out, voluntary programs have figured prominently in the national policies of many countries over the past decade and continue to play a leading role in Japan.

This prominence begs the obvious question of whether these programs are working. For example, following the 2002 announcement by President Bush of voluntary efforts to achieve an 18% improvement in greenhouse gas (GHG) intensity, recent data indicating that we are on track to meet that 18% goal have been cited as evidence of the voluntary programs' success (White House 2006). But are the data really evidence of voluntary programs' success, or do they simply reflect other coincident events? Equally important, as we move forward, what is reasonable to expect from such voluntary programs? We hope to contribute to the discussion of these questions through a careful analysis of the Climate Wise program. Although it is not the most recent voluntary initiative, and it did not benefit from potential improvements in voluntary program design that occurred in the late 1990s, the program's lengthier history makes it particularly amenable to statistical analysis.

Officially established by the U.S. Environmental Protection Agency (U.S. EPA) in 1993, Climate Wise was a voluntary program focusing on the non-utility industrial sector's efforts to encourage the reduction of carbon dioxide (CO_2) and other GHGs via adoption of energy-efficiency, renewable energy, and pollution-prevention technologies. Climate Wise remained in operation until 2000, when it was renamed and placed under EPA's Energy Star umbrella. Unlike Green Lights or EPA's other technology-based programs that require the adoption of particular technologies, Climate Wise members had the flexibility to use whatever technologies or strategies they chose to reduce their emissions. The basic requirements of Climate Wise were that a participating firm develop baseline emissions estimates of its GHGs for any year since 1990, pledge forward-looking emissions-reduction actions, and make periodic progress reports. As part of the program's operations, Climate Wise provided public recognition and certain types of technical assistance to its members. At its peak, Climate Wise had enrolled more than 600 industrial firms covering several thousand facilities nationwide. More recently, Climate Leaders, a program noted above with some design features similar to those of Climate Wise, has been embraced by the Bush administration as a key element of its climate change initiative.

There has been little outside evaluation of Climate Wise, with EPA estimates ranging from 3 to 20 million tons of emissions reductions. The principal goal of this chapter is to first review these existing discussions of Climate Wise accomplishments and then to try to tease out more accurate estimates of the program's impact using statistical techniques. As is well known, empirical studies examining the performance of voluntary programs are fraught with many difficulties, including the development of a credible control group or baseline against which to compare the performance of participating firms or facilities. As part of our efforts to develop a baseline, we are fortunate to have access to confidential plant-level data files for the manufacturing sector collected by the U.S. Census Bureau. Using this data, including information on industrial classification, geographic location, value of goods produced, energy expenditures, and—importantly—the cost of fuels, including CO_2-emitting fossil fuels, we develop a matched control group of plants against which to compare the performance (measured by growth in the cost of fuels) of observed Climate Wise participants. Our basic conclusion, with some caveats, is that Climate Wise had only a transient effect on fuel use and, therefore, emissions, with our best estimate being a 3% reduction.

Following this introduction, the next section presents background information on the Climate Wise program. Section three describes our methodology and data, including how the Climate Wise participant list has been linked to census data and how we have developed a sample of comparable firms. Section four presents empirical results, and section five offers a set of tentative conclusions.

Background

Following adoption of the United Nations Framework Convention on Climate Change in 1992, there was great interest in developing voluntary measures to meet one of the convention's stated goals of reducing GHG emissions to 1990 levels by the year 2000. In October 1993, President Clinton announced the Climate Change Action Plan to "meet the twin challenges of responding to the threat of global warming and strengthening the economy" (Clinton and Gore 1993). The view in the Clinton Administration was that Green Lights and other previously established programs held great promise and an ambitious effort was launched throughout the federal government to identify other opportunities for potential programs that could reduce emissions at low or no cost.

Unlike certain government-sponsored voluntary initiatives that were developed in the face of an impending threat of regulation, such as EPA's 33/50 program to control toxic releases, there was no realistic prospect for a mandatory control program for GHGs in 1993 when the Climate Wise program was established. Rather, Climate Wise was designed to appeal to the growing interest—arguably, self-interest—of many firms to reduce their energy costs and, at the same time, receive public recognition for reducing their emissions at a time when the climate change issue was coming into focus as a major public policy concern. A further advantage to participating firms, beyond any immediate cost savings or favorable publicity they might receive, is the knowledge they might gain about cost-effective options to reduce their GHG emissions should some future obligation occur and, perhaps, some influence in helping shape such an obligation. Arguably, if a firm had a better understanding of its own emissions and the likely benefits and costs from undertaking particular actions, it would be better prepared in the future to make the potentially large-scale emissions reductions that were being discussed in the context of the framework convention.

From EPA's perspective, development of the Climate Wise program was important for similar reasons. Without the near-term prospect of legal authority to regulate GHG emissions, the agency was nonetheless anxious to make some headway on the important and growing problem of climate change. Voluntary programs provided an opportunity for EPA to gain first-hand experience with the complexities of measuring and tracking emissions and of working with industry to understand the nature of the technical and economic challenges that might be encountered in any future mandatory program. Further, since many agency officials believed that significant GHG emissions reductions could be achieved at low or no cost, voluntary programs presented an opportunity to achieve at least limited emissions reductions in the near term and, simultaneously, to develop real-world evidence to support the contention that mitigation of emissions was a practical option for U.S. industrial firms. Such evidence, if it could be developed, might ease the way for future mandatory programs.

Unlike other voluntary programs developed at EPA in the early 1990s, Climate Wise was designed and operated (until 1999) by the agency's policy office, as opposed to various program offices. While this distinction may not be of great significance outside the agency, the flexible program design (described below) had particular appeal to the economics-oriented staff of the policy office, as opposed to the technology-oriented programs designed and operated by the air office. An interesting question is whether the extensive flexibility embodied in Climate Wise, including, for example, the absence of technology adoption or other specific performance obligations, had measurable effects, either positive or negative, on program performance.

Program Requirements

As stated in the 1998 *Progress Report* (U.S. EPA 1998) the four broad objectives of the Climate Wise Program were to:

- Encourage the immediate reduction of GHG emissions in the industrial sector through a comprehensive set of cost-effective actions;
- Change the way companies view and manage environmental performance by demonstrating the economic and productivity gains associated with "lean and clean" manufacturing;
- Foster innovation by allowing participants to identify the actions that make the most sense for their organization; and
- Develop productive and flexible partnerships within government and between government and industry.

Climate Wise consists of three interrelated components. First, the pledge component asks firms to commit to taking cost-effective, voluntary actions to reduce GHG emissions. Second, tailored assistance efforts are designed to facilitate companies' emissions-reducing efforts via a clearinghouse, workshops, and seminars. Finally, communication activities provide public recognition for actual progress in reducing emissions.

The requirements for a firm to participate in Climate Wise were relatively straightforward. To join, a firm had to develop a baseline estimate of its direct emissions of CO_2 (and other GHGs) for the year it joined the program or any prior year of its choice since 1990. Since an estimate of baseline emissions does not involve the detailed accounting information required for a full emissions inventory, the burden on the firm was relatively modest.[2]

In addition to establishing a baseline, a firm was required to identify specific actions it proposed to undertake to reduce its emissions and, for each action, to indicate whether this is a "new," "expanded," or "accelerated" initiative. To encourage the consideration of substantial reductions, EPA provided a checklist of major actions to improve equipment and processes, including those involving boiler efficiency, air compressor systems, steam traps, and piping and heat generating equipment. Also included were fuel-switching and best management practices, as well as the further integration of energy efficiency in new product design and manufacturing. Firms were strongly encouraged, albeit not required, to select at least some of their proposed actions from this list. The only formal requirement was for a firm to establish an emissions goal for the year 2000 and to provide a progress report directly to EPA. Participants also were encouraged, but again not required, to report their progress to the U.S. Department of Energy through the 1605(b) registry program.[3]

EPA provided several types of technical assistance to participating firms, including a guide to industrial energy efficiency, various government publications on energy efficiency and related issues and, most importantly, free phone consultation with government and private-sector energy experts retained as consultants by the agency. Information about financial assistance to support emissions-reducing actions also was made available to participants, including via Small Business Administration-guaranteed loans, low interest buy-downs from state providers, utility programs, and others. Further, EPA set up an annual event open to the public to recognize the performance of outstanding Climate Wise participants. As part of these events, a series of workshops were held that allowed participating firms to exchange experiences about their efforts to improve industrial efficiency and reduce GHG emissions. Informal reactions from agency staff and industry representatives suggested that these workshops were seen as quite valuable by the participating firms. A related effort involved encouraging Climate Wise participants to join other government-sponsored voluntary programs, such as Waste Wise and Motor Challenge.

Although the focus of the Climate Wise program was on energy efficiency and the reduction of CO_2 emissions, a number of firms did propose to reduce emissions of non-CO_2 GHGs as well. Reportedly, the most substantial reductions of the non-CO_2 gases were in the chemical industry, where relatively large amounts of nitrous oxide (N_2O) emissions were released in the manufacture of adipic acid. Significant amounts of methane also were included in the action plans of several firms, especially in the beer industry.

Over time, as the Climate Wise program evolved, EPA established a number of state and local programs under the Climate Wise logo. The purpose of these efforts was to help recruit additional participants and to further disseminate the growing body of Climate Wise technical materials that were being developed. Several state and local workshops were held to attract new Climate Wise members and to assist existing members in meeting their reduction targets. As noted, at its peak, Climate Wise counted more than 600 participating firms representing several thousand facilities across the country.

Program Accomplishments

The Climate Wise *Progress Report* issued in 1998 presents a detailed description of the program and its accomplishments. In 1997, for example, it claims that Climate Wise participants "started or completed more than 1,000 energy efficiency and emission reduction projects eliminating almost 3.3 million tons of carbon from the atmosphere" (*4*). An earlier agency report, *Pollution Prevention Accomplishments* (U.S. EPA 1994) claims even larger gains: "Eight Climate Wise companies, representing more than 3% of U.S. industrial energy use, committed to actions which will reduce GHG emissions by more than 20 million tons by the year 2000" (*17*). Interestingly, the 1998 *Progress Report* also claims that Climate Wise participants will save more than $240 million annually by the year 2000. Congressional testimony by a senior agency official in March 2000 claimed that Climate Wise participants representing 13% of U.S. industrial energy use had entered into agreements and submitted comprehensive plans to reduce their emissions. General Motors was cited as a specific success story:

> It has implemented a wide array of actions, including energy project competitions, more than 800 conservation projects, more than 370 energy partnership projects, and fuel switching for 5 steam generating powerhouses. These actions have resulted in emission reductions of about 2 million tons of carbon dioxide. (Stolpman 2000, 6)

At the same time EPA was claiming these accomplishments, a 1995 evaluation by James McCarthy of the Congressional Research Service pointed to the difficulty of judging the program effects. Specifically, he notes that:

> The amount of reduction projected is large; under the President's Climate Change Action Plan, it appears to represent about 20 percent of the *total* reduction of GHGs necessary to return U.S. emissions to 1990 levels by 2000. The baseline against which reductions are being measured is not specified in EPA's report, however; thus, it is not clear whether this comparison is valid.
>
> Further, the wide range of potential actions to be taken by the Climate Wise companies makes it unclear how significant a role EPA is playing in stimulating the company actions. That is, of course, a potential problem in judging the effectiveness of all of the voluntary programs. But in most of the others, the focus is sufficiently narrow that EPA's sponsorship of the program and the related publicity concerning its participants' accomplishments may be assumed to have some role in bringing about the resulting action. In this program, the potential actions are so diverse, and in many case so specific to production processes and

locations, that one may perhaps question whether EPA's role isn't more that of scorekeeper than catalyst." (McCarthy 1995, 6)

This criticism of EPA estimates points to exactly the same concern voiced throughout this volume: namely, that it can be exceedingly difficult to identify a meaningful baseline for evaluating the effects of a voluntary program. The criticism also draws out a related concern more specific to the flexible approach in Climate Wise: namely, that compared to more narrowly focused programs— say Green Lights or Energy Star—targeting a specific set of actions, it is harder to understand how a voluntary program like Climate Wise is catalyzing such widely varying activities. In either case, the key to a more objective and convincing evaluation is a transparent baseline against which to measure the behavior of program participants.

Methodology and Data

The underlying challenge noted in the previous section is to identify a suitable baseline against which we can evaluate the behavior of Climate Wise participants. In this section, we break that down into two pieces. First, we need a measurable outcome that we can observe both among participants and nonparticipants and that ought to reflect program effects. Second, we need a control group of nonparticipants that is sufficiently similar to the participants that it can be used to create a baseline.

Using detailed census data linked to the list of Climate Wise participants, our basic approach to both problems is to compare the growth in fuel costs after a facility enrolls in Climate Wise to the growth in fuel costs over the same period at a matched, non-participating facility. To add depth to our analysis, we consider different matched samples, different growth horizons, and additional outcome variables, including growth in the value of shipments (e.g., output) and electricity expenditures. The remainder of this section summarizes the methodology and the next section discusses results.

Difference-in-Differences Outcome Model

A key strength in the census data we use is that we have multiple observations over time of the same plant. This allows us to follow plants before and after they join Climate Wise, observing changes that might be attributable to the program. Among the data collected by the census surveys, the most relevant variable for evaluating Climate Wise is the one measuring cost of fuels (CF). This variable measures expenditures (e.g., price times quantity purchased) on all fuels—GHG-emitting fossil energy as well as wood, biomass, and other fuels. But because other fuels make up such a small portion of industrial energy use—in 2000, biomass and renewables comprised about 7% of primary energy consumption in the industrial sector (EIA 2006, *Table 2-1d*)—we feel this is a reasonable proxy for changes in fossil energy use and, therefore, emissions. We focus on the changes in the natural log of expenditures both to make the effects comparable across different size plants and to allow us to interpret the estimated differences as percent changes.

While this change over time in logged cost of fuels is a useful starting point, the problem with a before-and-after estimate among program participants is that there likely are many other changes unrelated to program participation that may be occurring at the same time and affecting the outcome variable. To control for these effects on the outcome variable that are unrelated to the program, we compare the before-and-after change at each participating plant to the change over the same period of time at a matched, non-participating plant. Letting $Y_{i,t}$ represent the cost of fuels at plant i and year t, we calculate:

$$\Delta_s Y_i = \left(\ln Y_{i,t(i)+s} - \ln Y_{i,t(i)-1} \right) - \left(\ln Y_{i',t(i)+s} - \ln Y_{i',t(i)-1} \right) \qquad (1)$$

for matched pairs of participant i and nonparticipant i', where $t(i)$ is the year in which participant i joined Climate Wise and s is the horizon over which we are examining the effect of the program. That is, we measure the change in logged cost of fuels starting the year before the participant joined Climate Wise $(t(i) - 1)$ and looking out over a horizon of 1, 2, or 3 years based on the value of s (e.g., year $t(i) + s$). We refer to $\Delta_s Y_i$ as the s-year horizon effect for observation i. With these difference-in-differences calculations in hand for each matched pair, we then calculate the overall program effect for all participants over horizon s, $\Delta_s Y$

$$\Delta_s Y = \frac{1}{N} \sum_{i=1}^{N} \Delta Y_i \qquad (2)$$

where N is the total number of matched participants. Here, the estimated error is easily calculated based on the standard deviation of the calculated ΔY's.

This difference-in-differences model—referring to the difference both in time and across participant and control—allows us to measure the program effect controlling for both the previous level of the outcome variable (the first difference) and changes in the outcome variable that are unrelated to the program (the second difference). The key challenge is to identify matched controls that indeed control for changes unrelated to the program; that is, controls that are similar to the program participants in every important way (as regards the outcome variable). We now turn to that question directly.

Matched Control Groups

It is useful to recognize that the justification for our approach is an assumption that the decision to participate is random and unrelated to any observable information once we have created our matched pairs. In other words, the participant and nonparticipant observations look the same for practical purposes except that some have randomly chosen to participate. In this way, when we look at outcome differences between the participants and controls, we know whatever differences we find are uncorrelated with—and essentially unrelated to—any other observable features and, therefore, can be ascribed to the program (this leaves open the possibility that observed differences might be related to unobservable features, an issue we return to later).

Our specific approach to matching is modeled after work by Rosenbaum and Rubin (1983) and more recently used by List et al. (2003) and Dehejia and Wahba (2002) that evaluates the effect of different programs based on a matched sample where participants and nonparticipants are similar. The general problem of creating a set of matched, non-participating observations is quite challenging: There are many observable variables—in our case describing location, industry, size, energy intensity, and growth—that we would want to match in order, arguably, to have controls that are similar to the participants. The important result based on Rosenbaum and Rubin (1983) is that we only need to match the expected likelihood of participation, rather than all of the key observable variables, for the difference-in-differences estimator to be unbiased. That is, we simplify the difficult problem of matching all these different variables to a much simpler one of matching a summary variable describing the propensity to join the program. This approach is referred to as propensity score matching.

Mechanically, we match program participants with suitable controls by first estimating a model predicting whether or not a plant joins the program in a given year. We then use that estimated model to predict the likelihood of participation, or propensity score. Finally, we take each participant i, find the nonparticipant with the closest propensity score, and call that observation the matched nonparticipant i'.

Data

To proceed with this approach, a list of voluntary program participants who joined Climate Wise in each of its operational years from 1994 to 2000 was obtained from EPA. This list includes name, zip code, and join date data for two different types of participants: those who joined at the corporate level and those who joined as individual plant participants. There were a total of 671 participants with complete data. In Table 7-1, we show the distribution of both types of participants over time. An interesting observation is that the number

TABLE 7-1. Join Data for Climate Wise Participants

Join year	Corporate	Plant	Subtotal
1994	8	0	8
1995	30	7	37
1996	141	38	179
1997	101	37	138
1998	70	36	106
1999	17	72	89
2000	0	114	144
Subtotal	367	304	701

of corporations joining reached a peak at 1996 and dropped to zero in 2000. However, the number of plant participants continued to increase until the program ended in 2000.

The information on program participation then was linked to detailed data from the Census Bureau's Longitudinal Research Database (LRD) using name and, for plant participants, zip code information. We succeeded in linking a total of 377 out of 671 participants, including 228 corporate participants and 149 plant participants. To some extent, the failure to link participants to the census data reflects the fact that census data only include manufacturing establishments, while the Climate Wise program includes both manufacturing and nonmanufacturing participants (e.g., municipalities, commercial buildings, etc.), despite its programmatic focus on manufacturing. Linking participants also may fail because names differ or because the participants were not included in the census survey.

The 377 linked participants from the original Climate Wise list translate into 2,311 facilities because corporate participants can have multiple associated facilities and, in a few instances, multiple facilities appear to match a single Climate Wise plant. These numbers are summarized in Table 7-2 (note the number of linked participants eventually will be halved due to missing yearly observations for particular plants as we begin to do our estimation and calculations).

TABLE 7-2. Matching of Climate Wise (CW) to Longitudinal Research Database (LRD)

	CW list	LRD plants	LRD plant-year observations (1992–2001)
Corporate participants with multiple plants	135	2,053	11,503
Corporate participants with a single plant	93	95	316
Plant-level participants	149	163	946
Total	377	2,311	12,765

TABLE 7-3. Sample Statistics, Longitudinal Research Database (LRD) and Program Participants

Variable	Summary Statistics	Full LRD sample (1992–2001)	Program participants
ln(TVS)	Mean	7.61	10.87
	Standard deviation	2.30	1.81
	Plant-year observations	1,157,606	12,605
ln(CF)	Mean	2.54	5.31
	Standard deviation	2.12	2.23
	Plant-year observations	839,934	11,280
ln(PE)	Mean	3.17	6.31
	Standard deviation	2.21	1.83
	Plant-year observations	1,019,042	12,377
	Number of plants	515,189	2,311

Notes: TVS = total value of shipments; CF = cost of fuels; PE = purchased electricity.

Summary statistics for the linked sample, as well as for the entire census database, are given in Table 7-3. Here, we see the principal differences among participants and the broader universe of plants in the census data: the participants are considerably larger. A difference of 2–3 units in logs translates into a factor of 10–20 in levels. Our participant sample also is a very small fraction of the plants in the census database—roughly 1%. This indicates that the full census sample is unlikely to be an appropriate control group as a whole and that there are a large number of plants from which to choose a more appropriate sub-group of controls.

It is worth noting that the pattern of joining Climate Wise (shown in Table 7-1) and consequent linking with census data turns out to have important consequences for our ability to evaluate the effect of program participation over longer horizons. As we try to look at behavior two or three years after participants joined the program, we are forced to drop plants that joined in 2000 and 1999, respectively, because our census data end in 2001. As noted, corporate participants provide the overwhelming majority of participant observations because they match to multiple facilities. Given the steep drop-off in new corporate participants after 1998, we do not sacrifice many observations by looking two to three years out. However, trying to discern effects four years after joining, with only participants who joined between 1994 and 1997, we have noticeably fewer observations and noticeably noisier estimates. For this reason, and the later observation that effects appear to vanish after three years, we do not attempt to look at effects more than three years after participants join the program.

Estimation and Results

Our evaluation strategy has two parts: creating matched controls based on an estimated model of program participation and then creating difference-in-differences estimates of the outcome. The simplicity of the difference-in-differences estimation provides little opportunity for variety. We can examine other variables (namely value of shipments and electricity expenditures, as well as our primary variable of interest, cost of fuels) and we can examine different horizons (e.g., effects one, two, or three years after joining the program) but that is the extent of possible variations.[4] The participation model, however, potentially is more elaborate. Participants join in different years, raising the issue of whether there should be a single participation model or different models for different years. There also are an array of control variables and combinations of control variables: starting values for cost of fuels, electricity expenditures, value of shipments, industry and regional controls, and various measures of growth. In the remainder of this section, we first describe the estimated participation models and then present estimates of program effects.

Propensity Score Matching

Even before we begin specifying the participation model, we have to decide whether one model will govern the participation choice across all years or whether we will estimate different models in different years. Considering Table 7-1, we know that different models in different years will be problematic because there are few observations in later years; therefore, we have chosen to estimate a single model. Specifically, we estimate a Cox hazard model: In each year, there is a baseline probability of joining Climate Wise for all plants that have not joined in a previous year. Different values of various independent variables affect the relative probability of joining versus that baseline probability. The Cox model imposes no restrictions on the pattern of baseline hazard over time. We then consider alternative specifications of the independent variables predicting the relative hazard.

We focus on three sets of independent variables. We first include variables describing the size of each plant and energy use at each plant. That is, variables describing the value of shipments (TVS), cost of fuels (CF), and electricity expenditures (EE) in the previous period. Here, we use data from the preceding period because we do not know when during the year plants join and we want to be certain that we are measuring the effect of these variables on the propensity to join, not the other way around. We consider both linear and quadratic (squared and interaction) terms. Second, we include dummy variables indicating manufacturing sector (roughly 20 categories) and region of the country (9 categories). Third, we include a variable reflecting the future growth of each plant measured as the change in logged value of shipments one, two, or three years in the future versus the previous year. We choose the horizon of growth to match the horizon over which the program effects are being measured (e.g., when we are looking at the effect of Climate Wise participation after three years, we use a propensity model with a three-year growth variable to match the data).

The need to include a growth variable became evident when we observed, after matching only on the other two sets of variables, that participants grew faster than nonparticipants. We did not believe that joining the program caused faster growth but rather that faster-growing firms were more likely to join. That would be consistent with both our and other researchers' observations that larger plants are more likely to join (e.g., the raw statistics in Table 7-4). This leads to a mathematical model of the form:

$$\begin{matrix} \text{probability of joining in year } t \\ \text{(assuming plant } i \text{ has not yet joined)} \end{matrix} = h(t)\exp\begin{pmatrix} \beta_{size}\ln TVS_{i,t-1} + \beta_{elec}\ln EE_{i,t-1} + \beta_{fuels}\ln CF_{i,t-1} \\ + \left[\text{all quadratic combinations of size, elec, fuels}\right] \\ + \beta_{growth}\left(\ln TVS_{i,t+h} - \ln TVS_{i,t-1}\right) \\ + \sum_{\text{industries } j}\beta_j 1\left(M_i = j\right) + \sum_{\text{region } k}\beta_k 1\left(G_i = k\right) \end{pmatrix} \quad (3)$$

where *TVS* is the total value of shipments, *EE* is electricity expenditures, *CF* is cost of fuels (all subscripted according to plant i and year t), M_i refers to the industrial classification of plant i, and G_i refers to the census region where plant i is located. The β's are parameters to be estimated. The expression indicates that the probability of joining in year t is conditional on not having joined previously equals a baseline probability of joining in year t, $h(t)$, followed by the relative hazard influenced by the variables described above.

With roughly 40 right-hand-side variables, we were concerned that our model might not be effectively capturing the influence of some of the more important variables. Specifically, if we do not accurately match for growth and participants are growing faster than the matched controls, we easily can imagine that this will bias estimates of any changes in cost of fuels or electricity to be more positive (versus our intuition that those changes will be negative, assuming program participation leads to reductions in energy use). For this reason, we considered models without the industry and regional dummy variables, as well as without the quadratic terms.

Estimated parameter values (β's in the equation above) for the one-year horizon models with various combinations of dummy variables and quadratic terms are reported in Table 7-4, except for coefficients on regional and industry dummy variables (that could not be released because of the confidentiality rules imposed by the Census Bureau). Note that the variables measuring size, electricity, and fuel use are highly correlated, making it difficult to estimate precisely coefficients on each. Overall, larger facilities are more likely to join the program (most easily seen by the coefficient on value of shipments in the lin-

TABLE 7-4. Parameter Estimates from Hazard Model Predicting the Probability of Joining Climate Wise

ln(TVS)	Full model	Linear terms only	No regional / industry effects	Linear terms only and no regional / industry effects
ln(TVS)	-0.317	0.331**	-0.704**	0.289**
	(0.285)	(0.038)	(0.251)	(0.034)
ln(TVS)²	0.020		0.046**	
	(0.021)		(0.018)	
ln(CF)	0.386**	-0.027	0.770**	-0.012
	(0.169)	(0.026)	(0.147)	(0.023)
ln(CF)²	0.021**		0.025**	
	(0.010)		(0.009)	
ln(EE)	0.024	0.251**	-0.081	0.244**
	(0.249)	(0.039)	(0.239)	(0.037)
ln(EE)²	-0.003		0.002	
	(0.021)		(0.020)	
ln(TVS) ln(CF)	-0.028		-0.070**	
	(0.021)		(0.018)	
ln(TVS) ln(EE)	0.049		0.047	
	(0.038)		(0.035)	
ln(EE) ln(CF)	-0.047**		-0.039*	
	(0.021)		(0.021)	
Growth in TVS[a]	0.157**	0.174**	0.164**	0.195**
	(0.078)	(0.076)	0.078	(0.076)
Growth²	-0.023		-0.043	
	(0.040)		0.039	
Participant count	1,024	1,024	1,024	1,024
Non-participant count	65,008	65,008	65,008	65,008
Log-likelihood	-9866.24	-9875.49	-10066.3	-10087.3

Notes: parameter estimates indicate whether variable increased or decreased the probability of joining; standard errors are in parentheses. TVS = total value of shipments; CF = cost of fuels; EE = electricity expenditures; ** = significant at the 99% confidence level.

a. The results in this table reflect growth in TVS over a two-year window, measured from the year prior, to the year after the current year. Similar results arise with longer windows. The predicted probabilities used to create matched samples are estimated with a growth window consistent with the horizon of the estimated program effect. That is, estimates of effects reported in Table 7-5 one year after joining are based on the above models; estimates of effects two years after joining are based on an estimated participation model (not reported) with a growth window extending two years past the current period, etc.

ear model). This is consistent with the observation in the data section that participants are, on average, considerably larger than other plants in the census sample. (Note that the inclusion or exclusion of dummy variables has little effect on the parameter estimates.)

The principal use of the parameter estimates in Table 7-4 is that they allow us to compute propensity values for both the participants and nonparticipants. Once estimated, we systematically consider each participant, identifying the nonparticipant with the nearest propensity score in that year. This constitutes our matched sample.

Difference-in-Differences Results

With the matched sample in hand, we compute the difference-in-differences estimates described in Equation 1. We do this for distinct matched samples created using the four participation models noted above (with and without dummy variables; with and without quadratic terms) and over one-, two-, and three-year horizons. That is, we have 12 matched samples (four models by three horizons) from which to calculate estimates. Finally, we report the results not only for our variable of interest—the cost of fuels—but also for total value of shipments (to test our effort to match on growth) and electricity expenditures. The results are compiled in Table 7-5.

The first three rows of results show that having included the growth in total value of shipments in the participation equation, the matched sample has successfully matched on growth in this variable. There are no statistically significant differences between the Climate Wise participants and the controls and no real pattern of positive or negative results across models or horizons. In particular, our earlier concern about the large number of variables in the model (and the possibility that this might lead to a poor match on growth) appears unwarranted. Skipping for a moment to the results for electricity expenditures, we see statistically significant and positive results over the one- to two-year horizon. There is an initial increase of 4–6% in electricity expenditures associated with joining Climate Wise, depending on the model, which vanishes after three years. This result is robust across models—something we return to below.

Our principal focus is whether Climate Wise had any measurable effect on the cost of fuels—that is, the purchases of fossil fuels leading to emissions of carbon dioxide. Here, in the middle three rows, we see mixed and insignificant results. The one- and three-year horizons suggest negative effects; the two-year horizon positive effects. The model with the fewest controls in the participation model (column 4) shows the least positive effect on the two-year horizon but the largest negative effect after three years. In general, the standard errors are larger for cost of fuels than for the other two variables (3–5%, versus 2–3%). Given all this, we are tempted to conclude that the program had no perceptible effect on fuel use.

Summarizing our initial observations, we appear to have successfully controlled for growth but find only insignificant results for fuel expenditures and a positive effect on electricity use associated with joining Climate Wise. That effect vanishes after a few years. How can we explain this pattern of results? It

TABLE 7-5. Difference-in-Differences Estimate of Voluntary Program Effect

	Alternate propensity models			
	Model 1	*Model 2*	*Model 3*	*Model 4*
Propensity model:				
Quadratic terms	x		x	
Regional and industry dummies	x	x		
Total value of shipments (TVS)				
1-year horizon	−1.3	−0.3	−2.3	2.4
	(1.8)	(1.8)	(1.9)	(1.8)
2-year horizon	−0.2	2.5	0.4	−2.1
	(1.9)	(2.0)	(2.0)	(2.0)
3-year horizon	0.9	−0.3	−0.1	0.9
	(2.2)	(2.5)	(2.3)	(2.4)
Cost of fuels (CF)				
1-year horizon	−3.3	−2.2	−3.6	−2.7
	(3.4)	(3.3)	(3.3)	(3.3)
2-year horizon	3.4	3.0	2.6	0.2
	(3.9)	(3.7)	(3.8)	(3.9)
3-year horizon	−2.0	−4.3	−5.1	−6.9
	(4.3)	(4.6)	(4.4)	(4.3)
Electricity purchase (EE)				
1-year horizon	5.0**	5.8**	3.6	6.6**
	(2.0)	(2.2)	(1.9)	(2.1)
2-year horizon	3.0	3.8	2.9	4.8**
	(2.4)	(2.2)	(2.5)	(2.5)
3-year horizon	0.0	0.5	2.5	1.2
	(2.7)	(2.8)	(2.6)	(2.6)

Note: Percent change in indicated variable, relative to year prior to joining, relative to matched control. Controls were matched on propensity scores estimated as a function of the indicated dummy variables, lagged values of TVS, CF, and EE, quadratic terms (including interactions) where indicated, and growth in TVS over the indicated horizon (and growth squared as indicated). All variables were measured in natural logs.
** Significant at the 5% level.

is one thing if Climate Wise has an imperceptible effect on the cost of fuels but it is quite a different situation if the program led to an increase in electricity use. Several possible explanations suggest themselves.

First, participating plants may have pursued emissions reductions that required increased electricity use. Ignoring the indirect emissions associated with electricity use, this technically reduces emissions as defined by the program goals but with the unintended consequence of higher indirect emissions from electricity use. Here, we might imagine that there was a small negative effect on fuel use but that it has vanished in the noise.

Second, the estimated effect on the cost of fuels may not reflect the actual effect on emissions. Fuel switching among purchased fuels—for example, a shift to biomass or from coal to gas—might reduce emissions without changing expenditures. Further, some plants may have pursued non-energy-related emissions reductions, such as N_2O emissions at chemical plants. In this way, there may be a negative effect on emissions that is not reflected in a lower cost of fuels.

Third, the effect on electricity expenditures may reflect a failure to adequately control for growth, despite the lack of such evidence based on the estimated effects on total value of shipments in the first three rows of Table 7-5. For example, even though growth in the value of shipments is negligibly different between participants and controls, we have no way of knowing about the underlying prices and quantity changes. Participants might experience changes in quantities while those matched from the census database might experience changes in prices; we cannot tease out controls that have that same pattern because there is no available detail on prices and quantities. If the estimated electricity expenditure growth effect really is reflecting an underlying and uncorrected difference in growth between participants and controls, then (presumably) fixing it would raise the growth rate of the control group and make the estimated program effect on electricity and fuel costs more negative.

Put another way, this relates back to our earlier observation that the approach taken in this analysis matches on observable variables. If there are unobserved differences, such as some plants increasing the quantity of output and others increasing prices, then even as the total value of shipments (or cost of fuels) appears the same among participants and the pool of controls, that similarity might be the result of a true program effect and a countervailing effect arising from unobserved differences in growth.

Regardless of how we explain the unusual pattern of program effects, a key observation is that all of the evidence suggests that these effects vanish after three years. That is, even if there is a perverse effect on electricity usage or even if a better model would show a decline in fuel use, the only statistically significant differences disappear over time. This is not a consequence of broadening standard errors and greater noise but a clear diminution in magnitude of estimated differences. In this way, it may be fair to say that the effect of Climate Wise was to accelerate energy-saving behavior that eventually arose among program participants and nonparticipants alike. Given that effects vanish after several years, coupled with concern over whether growth was adequately controlled (a problem that should only increase over time) and the particularly counterintuitive two-year horizon results for cost of fuels, we find the one-year horizon results to be the most convincing estimate of the transient effect on fossil fuel use. That is, as a best guess we would conclude that the program had a temporary, negative 3% effect on fuel use—a number that might be underestimated based on the observed 5% increase in electricity use. In any case, the standard errors alone suggest this could be off by ± 7%.

Conclusion

A key element of the U.S. policy response to climate change has been the adoption of voluntary programs, such as Climate Wise, to encourage companies to reduce GHG emissions. While voluntary programs have evolved and undoubtedly improved over the past decade—increasing the requirements, scrutiny, and technical assistance given to participants—they still form the core of the U.S. policy response. Yet, careful evaluation of these programs has been limited. Reports published on Climate Wise by EPA have provided a range of estimated effects, ranging from 3 to 20 million tons of reductions, while critics have asked whether these reductions are based on a realistic baseline of what would have occurred absent the Climate Wise program. This criticism has noted specifically the broad range of activities that were included in Climate Wise, suggesting that it is hard to understand how such disparate actions really could be causally linked to the program.

In this chapter, we discussed our efforts to link information on Climate Wise participation to detailed census data collected in the 1990s and then to evaluate participant behavior against a matched set of controls. Although a number of important caveats apply, our principal result is that Climate Wise at best had a temporary (one- to two-year) effect on participant behavior, increasing electricity expenditures and perhaps slightly decreasing fuel expenditures. We posit three explanations for the unexpected and statistically significant effect on electricity. First, it may legitimately reflect efforts that temporarily reduce fuel use and, hence, emissions, but do so at the expense of increased electricity use. Second, larger emissions reductions may not be reflected in the fuel expenditure variable, either because there was a shift to less-emitting fuels (gas and biomass) or because there were reductions in non-fuel emissions. Third, despite our efforts to match on observed variables, unmatched differences in unobserved variables may be contaminating the results. For example, the matched growth in the value of shipments might be masking higher growth among participants in the quantity of shipments, therefore explaining the growth in electricity expenditures and positively biasing estimates of the program's effect on fuel expenditures.

Regardless of these explanations, a key observation is that the effects appear to vanish after one or two years. For this reason, as well as concern that unobserved differences might be growing over time and that the two-year results are particularly counterintuitive, we tend to put greater emphasis on the estimated one-year effects. These indicate a statistically insignificant 3% decline in fuel expenditures and a 5% increase in electricity use. But more importantly, whatever the explanation, there is no evidence of a persistent effect after two to three years.

Apart from these empirical estimates, observations, and explanations, legitimate questions remain about the interpretation and application of our results more broadly. Climate Wise was a particularly broad, flexible voluntary program, whereas many programs are more specific and technology focused. We believe it is inappropriate to extend our results to those types of programs. But is it reasonable to interpret our findings as undermining claims of environmental effectiveness of other broad, flexible, performance-based voluntary programs that do not focus on specific technologies? We believe not. Climate Wise was the first flexible, nontechnology-based voluntary GHG reduction program introduced by EPA, and newer programs contain somewhat different design elements. Unlike Climate Wise, which had particularly modest requirements for participation and for the reporting of energy/environmental outcomes, the recent Climate Leaders initiative is more rigorous. For example, Climate Leaders specifically requires that participants develop emissions inventories through established protocols (as opposed to simple baseline estimates). It also has more rigorous goal-setting and reporting requirements. Whether these changes will have an effect on corporate behavior is unknown at this time. More stringent requirements might reduce the willingness of firms to join the new program. At the same time, it is possible that these new requirements will bring greater rigor and accountability to the initiative and lead to an increase in its environmental effectiveness.

Our work does allow us to conclude that self-reported results without an objective baseline can be misleading. Even if participants genuinely believe that a voluntary program is influencing their behavior and can point to actions they attribute to the program, the only legitimate benchmark is what other, similar, non-participating facilities are doing at the same time. While this chapter has highlighted obstacles to such an approach, we believe it still provides a better window into the accomplishments of voluntary programs.

References

Clinton, William J., and Albert Gore, Jr. 1993. *The Climate Change Action Plan.* http://www.gcrio.org/USCCAP/toc.html (accessed August 1, 2006).

Dehejia, Rajeev H., and Sadek Wahba. 2002. Propensity Score-Matching Methods for Nonexperimental Causal Studies. *The Review of Economics and Statistics* 84(1): 151–162.

EIA (Energy Information Administration). 2006. *Annual Energy Review 2005.* Washington, DC: EIA.

List, John A., et al. 2003. Effects of Environmental Regulations on Manufacturing Plant Births: Evidence from a Propensity Score Matching Estimator. *Review of Economics and Statistics* 85(4): 944–952.

McCarthy, James E. 1995. Voluntary Programs to Reduce Pollution. Congressional Research Service Report 95-817 ENR Washington, DC, July 13.

Rosenbaum, P., and D. Rubin. 1983. The Central Role of the Propensity Score in Observational Studies for Causal Effects. *Biometrika* 70: 41–55.

Rubin, D. 1974. Estimating Causal Effects of Treatments. *Journal of Educational Psychology* 66: 688–701.

Stolpman, Paul. 2000. Testimony before the Subcommittee on Energy and Environment of the Committee on Science. U.S. House of Representatives, Washington, DC, March 9.

U.S. EPA (U.S. Environmental Protection Agency). 1994. *Pollution Prevention Accomplishments*, EPA Report 100-R-95-001. Washington, DC: U.S. EPA.

————. 1998. *Climate Wise: Progress Report*, EPA Report 231-R-98-015, Office of Policy. Washington, DC: U.S. EPA.

White House. 2002. Global Climate Change Policy Book. http://www.whitehouse.gov/news/releases/2005/05/20050518-4.html (accessed August 1, 2006).

White House. 2005. Climate Change Fact Sheet. http://www.whitehouse.gov/news/releases/2005/05/20050518-4.html (accessed August 1, 2006).

White House. 2006. Ask the White House April 21, 2006. http://www.whitehouse.gov/ask/20060421.html (accessed August 1, 2006).

Notes

1. The authors are Senior Fellows, and Fellow, respectively, at Resources for the Future. The research in this chapter was conducted while the authors were Special Sworn Status researchers of the U.S. Census Bureau at the Center for Economic Studies. Research results and conclusions expressed are those of the authors and do not necessarily reflect the views of the Census Bureau. This chapter has been screened to ensure that no confidential data are revealed. The authors wish to thank Arnie Reznek and Javier Miranda for generous technical assistance. This work was supported by the Smith Richardson Foundation and a STAR grant from the U.S. EPA.

2. The Climate Leaders Program established by the Bush Administration in 2002 does require an emissions inventory.

3. Created under section 1605(b) of the 1992 Energy Policy Act, this registry allows companies to report voluntary emissions reductions activities in a standardized format. See http://www.eia.doe.gov/oiaf/1605/frntvrgg.html (accessed November 27, 2006).

4. One extension we do not consider is whether and how the estimated difference-in-difference varies by propensity. Our interest in the average effect among participants makes this a less interesting question.

8

The Evaluation of Residential Utility Demand-Side Management Programs in California

Alan H. Sanstad[1]

Over the past three decades, the state of California has developed and imple-mented an extensive regulatory infrastructure to mitigate the environmental consequences of energy use. A core goal of the state's energy–environmental policy paradigm is to stimulate the adoption of tech-nologies and practices to increase the end-use efficiency of electricity and natural gas consumption in the provision of energy services across the sectors of the economy. In turn, one of the most important means of reaching this goal is electric and gas utility-based demand-side management (DSM), which com-prises a range of programs and measures that directly promote conservation practices and energy-efficiency investments on the part of residential, com-mercial, and industrial customers.

This chapter discusses the evaluation of one type of DSM program: residen-tial energy management services (REMS). REMS programs offer to voluntarily participating customers detailed analyses of their particular energy consump-tion patterns, equipment holdings, and other relevant household characteristics and offer household-specific advice on conservation practices and potential efficiency investments. These are much more specific, and serve a different pur-pose, than mass-market informational programs that promote energy efficiency and conservation. Unlike most DSM programs, however, REMS programs do not offer financial incentives or subsidies for the purchase of energy-efficient equipment. Moreover, like all types of DSM programs, REMS programs can be contrasted with the direct regulation of equipment and building energy use through minimum efficiency codes and standards, a regulatory approach that has been concurrently applied in California, other U.S. states, and at the national level over the last 30 years. REMS programs are therefore in a certain sense canonical voluntary programs in the energy arena. They are thus of interest from an evaluation perspective despite being known to achieve, as would be expected, generally lower energy savings than programs that offer financial incentives.

From the inception of DSM in the 1970s and for several decades following, both resources allocated to these types of programs and the resulting estimated energy savings grew in California. Following something of a de-emphasis during the latter half of the 1990s, the resurgence of attention to policy problems associated with energy and the environment (particularly, in California, electric power supply reliability and global climate change) has resulted in a renewed emphasis on DSM and other approaches to directly stimulating the adoption of energy-efficient technology and conservation practices.

Unlike many voluntary environmental programs, REMS and other forms of DSM are not programs per se but approaches to achieving energy efficiency and conservation goals that are instantiated in many individual programs. Thus, this chapter examines evaluation on two levels, both considering several specific programmatic examples and tracing the institutional and policy history of DSM and its evaluation in California. These foci are complementary. Assessing the effectiveness of DSM, of both REMS and other types of programs, ultimately rests upon measurements of the outcomes of individual programs. At the same time, ensuring that such measurements are obtained systematically and using appropriate methods is an institutional problem. The manner in which this particular institutional capacity has evolved in California over the past several decades is not widely known outside the evaluation community but is important for judging the past performance and future potential of programs of this type. This is especially the case given the leadership role played by the state in environmental and energy policy.

The chapter begins with a sketch of the origins and evolution of and rationales for utility DSM and other technology-oriented energy and environmental programs and policies in California. It then reviews estimates of the effects of these programs and discusses two central issues in their evaluation: the accurate ex post measurement of energy savings and the free-rider problem—the presence of program participants who would have undertaken program-recommended actions, or would have been inclined to, in the absence of the program and its statistical manifestation, the self-selection problem in quantitatively modeling and estimating program effects.

The development in the early-1990s of California's first regulatory framework for DSM evaluation is next summarized, followed by a review and discussion of evaluations of three specific REMS programs in the 1990s, with an emphasis on the treatment of the selection problem. The chapter then sketches more recent developments in methods for addressing selection bias and ends with summary remarks on the findings.

Origins of and Rationale for
Utility Demand-Side Management

DSM has its origins in an engineering and technology-based, or bottom-up, approach to analyzing technical and policy-related aspects of end-use energy consumption, efficiency, and conservation that emerged in the late 1960s and early 1970s amid the rapid expansion of environmental regulation in the United States and of environmentalism as a political and intellectual movement. This paradigm was first systematically articulated in a report of a meeting of the American Physical Society in 1974 on the physics and other technical aspects of energy consumption from an end-use perspective (Carnahan et al. 1975). The fundamental principle was the application of the second law of thermodynamics to distinguish between energy—that is, fuel—use, and the production of energy services such as space conditioning, lighting, refrigeration, and motor vehicle transportation. Focusing on the efficiency with which fuel inputs were transformed into useful energy services opened the way to making this technical nexus a foundation for the analysis of energy demand and the formation of public policies directed at the demand side of markets for energy and energy-using technologies.

Both technical and policy analysis based on these principles expanded rapidly in California in the 1970s, and the end-use efficiency paradigm has remained a cornerstone of California energy policy since that time.[2] On the technical side, the University of California at Berkeley and the Lawrence Berkeley National Laboratory, among other institutions, became centers for this form of analysis. On the policy side, in 1974 the California State Legislature passed and then-governor Reagan signed into law the Warren-Alquist Act, which established a significant broadening of the declared public interest in the production and consumption of energy based on introducing environmental considerations, including air quality. Environmental and other social criteria were introduced as primary goals of publicly regulated energy utilities, which were instructed to promote conservation and efficiency in addition to their hitherto exclusive emphasis on developing energy supply resources, as well as to begin developing renewable supply resources such as wind, solar, and geothermal power. These goals also were motivated by a quickly emerging concern regarding the reliability of energy supplies in the wake of the oil embargo of 1973.

The act also put into place strict new guidelines regarding the planning for and construction of power plants; although not explicitly stated, a key motivation for the legislation was to forestall a major shift of the state's electric power system toward nuclear generation. The act also created the California Energy Commission (CEC) as the public agency responsible, in cooperation with the California Public Utilities Commission, for overseeing such efforts and ensuring the inclusion of environmental criteria in the provision of energy and energy services in the state. Among the CEC's specific areas of authority were the development and implementation of minimum energy-efficiency performance standards for appliances and buildings; such regulations went into effect in the state in the mid-to-late 1970s.

In the following years, policies and programs to affect directly utility customers' choices regarding energy utilization and the adoption of energy-efficient technologies—that is, demand-side management—saw initial development and application. The official beginnings of DSM can be traced to the National Energy Policy Act of 1978, one of the components of which, the National Energy Conservation Policy Act, instructed utilities to begin offering energy conservation services to customers to slow the growth of electricity demand. Data of the CEC, however, record small energy savings from utility efficiency programs of this type in California as early as 1976. By the late 1970s, the California Public Utilities Commission (CPUC) had instituted least-cost planning, a formal protocol for comparing investments in efficiency with those in new generation. Utility DSM was a mechanism whereby efficiency could be promoted when it was favored in this kind of comparison.

As it has evolved, DSM has come to encompass several types of programs to promote the adoption of energy-efficient technology and energy-conserving behavioral practices, as well as direct load-control interventions (such as incentives to shift electricity use to off-peak hours).[3] Informational DSM programs include both those that convey general information about energy efficiency and conservation (for example, through mass media campaigns) and site or household-specific technical information, particularly energy audits, in which individual customers' energy-use patterns are analyzed in detail and specific recommendations are provided for energy-saving and cost-reducing investments in equipment or thermal shell measures and for changes in practices, such as in the utilization of existing appliances. The larger category of DSM, in all sectors to which it has been applied, gives financial assistance for efficiency investments in the form of subsidized loans, direct payments, or simply the installation of equipment at no direct cost to the customer. DSM for demand or load management offers incentives for shifting the timing of electricity use and/or the installation of equipment, such as programmable thermostats, that can accomplish such shifting automatically, as well as real-time pricing or other changes to tariffs to induce load shifting.

Although not directly relevant to the technical issues of program energy-savings estimation, it is worth remarking on a long-running debate regarding the economic underpinnings of DSM and other technology-oriented energy-efficiency policies and programs. While this debate has been less vigorous in the case of purely informational programs such as REMS (in contrast to both efficiency performance standards and DSM programs in which explicit incentives are paid), it has nevertheless colored the manner in which they have been viewed in the energy-policy arena.

It was found very early on in the history of DSM that expert estimates of the cost-effectiveness of specific energy-efficient technologies or measures often were not reflected in the decisions of adopters, whether households or firms. The characteristic indicator of this divergence revealed hurdle rates for efficiency investments that appear to exceed substantially risk-adjusted market interest rates.[4] This phenomenon led to the introduction of the idea of market "barriers" to the adoption of efficient technology, such as insufficient information, lack of financing, differences between technology purchasers and users (the landlord-tenant problem), and others (Blumstein et al. 1980). That these barriers for the most part did not coincide with the market failures recognized by neoclassical welfare economics led many, if not most, energy economists to conclude that DSM and other technology-oriented policies and measures to promote energy efficiency were aimed at a problem that did not obviously exist in the first place. Economists' rebuttable presumption for the most part has been that the underlying environmental goals toward which technology-focused policy instruments are directed would be better reached through incentives in the form of, for example, emissions taxes—that is, through the internalization of externalities through the price mechanism. [5]

Sanstad et al. (2006) observe that, regardless of the putative economic superiority of price-based instruments for environmental protection, explicit reflection of environmental costs in energy prices has to date been politically infeasible in the United States. Thus, in practical policy terms, the actual choice has not been in general between price-based and technology-based policies to improve energy-efficiency as a means of achieving environmental goals but rather between technology policies and no policies at all. From this point of view, even if price instruments would be optimal, nevertheless DSM, efficiency performance standards, and other bottom-up approaches can be considered as feasible second-best alternatives.[6]

In any case, perhaps surprisingly and certainly unfortunately, this debate is little closer to resolution than it was when it originated.[7] It has, however, had little practical effect on the evolution, or continued popularity, of technology-oriented policies and programs in California, although the issues involved may become more salient in the context of the state's new greenhouse-gas mitigation targets and the emphasis on expanded energy-efficiency policies and programs as one of the means to reach them.[8]

DSM Program Evaluation in California

Along with other efficiency-promoting policies and programs, utility-based DSM in California grew steadily although somewhat cyclically following its emergence in the 1970s. Expenditures on these programs reached nearly $150 million by 1984; following a several-year decline, funding rose to approx-

TABLE 8-1. Electricity Savings from Energy-Efficiency Policies and Programs in California, 1980–2000

	1980	1990	2000
Statewide electricity consumption	166,976	228,038	263,599
Savings from all policies and programs	5,145	22,988	35,853
	(3.1%)	(10.1%)	(13.6%)
Savings from utility energy efficiency	3,692	13,091	16,757
and conservation programs	(2.2%)	(5.7%)	(6.4%)
Savings from building codes and	1,288	8,186	17,641
appliance standards	(0.8%)	(3.6%)	(6.7%)

Note: Electricity consumption and savings in GWh, percentage savings as percent of total consumption.
Source: CEC (2003a).

imately $275 million in 1994, and following another trough to $300 million in 2000 (CEC 2003b). CEC estimates of the electricity demand reductions from DSM and other efficiency policies and programs are illustrated in Table 8-1; although the available data do not distinguish the purely voluntary type of DSM from incentive programs, they do convey the general magnitudes involved (CEC 2003a).

All types of utility conservation and efficiency programs are estimated to have saved an amount equivalent to about 2% of total statewide electricity consumption in 1980, about 5.5% in 1990, and about 6.5% in 2000, with all demand-side programs combined saving about 3, 10, and 14% in these years, respectively, with nearly all the savings beyond the conservation and efficiency programs resulting from building codes and standards and appliance standards.[9] Note that these figures in each year represent cumulative savings; that is, savings both from programs initiated in that year and continuing savings from programs initiated in previous years.

Results of a review of California DSM evaluation studies illustrate residential program outcomes circa 1990 (Brown and Mihlmester 1994). In the years 1990 to 1992, the four investor-owned utilities in California were reported to have conducted 25 residential DSM programs serving approximately 813,000 customers. Of these, 5 programs serving about 222,000 customers were of the REMS type. Savings from the latter were not reported separately, but first-year savings from 9 incentive measures were estimated to have been about 118,000 megawatt-hours (mWh). More generally, REMS programs long have been recognized as delivering lower levels of energy savings than their companion programs offering financial incentives; reviewing nationwide program results through the 1990s, Nadel (1992) offers the estimate of 0–2% of baseline consumption for mass-market information programs and 3–5% for REMS-type programs.

Estimates of this type are calculated primarily on the basis of utility data and are therefore subject to uncertainty arising from a variety of factors, perhaps most importantly the changing character of the criteria for evaluating and measuring program effects over several decades.[10] At the program and regulatory levels, formal procedures for evaluating DSM programs emerged over a number of years.[11] In the first decade or so following the inception of this approach in the mid-1970s, such procedures were generally lacking. A common if not universal approach to "evaluation" was to compute and report ex ante engineering-based estimates of savings that would result from specific programs. In the case of residential DSM, shortcomings in evaluation methodologies became apparent by the early 1980s, in part due to problems with savings estimates for the Residential Conservation Service (RCS), a federally mandated program based in the 1978 National Energy Conservation Policy Act. RCS was a program based on home energy audits—detailed assessments by utilities of the energy demand characteristics of individual households that could be used to provide efficiency investment and conservation recommendations. It also included such measures as federal tax credits for efficiency investments, subsidies for efficiency and conservation in low-income households, and support for loans for conservation measures in federal secondary mortgage markets. As detailed in Walker et al. (1985), weaknesses in RCS evaluations were significant and widely recognized; although a complex set of issues was involved, the verification of savings was a central problem. In part as a result of these evaluation problems, there were great uncertainties in the overall results of the program, both in energy savings and in cost-effectiveness. At the same time, the considerable effort invested in, and experience gained from, evaluation of RCS programs contributed substantially to the subsequent development of DSM evaluation.

The importance of improving DSM evaluation methods and practices became increasingly clear to California regulators during the course of the 1980s. Further practical experience with the programs indicated that savings estimates frequently, if not characteristically, overstated programmatic effectiveness. With the decision in the 1980s by regulators in California and elsewhere to begin granting utility shareholders incentives based on the performance of DSM programs, this became a major issue in both utility regulation and DSM program planning and practice (NARUC 1989).

While many measurement, modeling, and other issues were and are involved in obtaining reliable estimates of program effects, much of the effort to establish a structure for more reliable and robust evaluation centered on two underlying problems. First was that of knowing what actual effects the programs were having; that is, the need to measure savings ex post rather than rely solely on ex ante estimates. The term "realization rate" was coined, defined as the ratio of measured, ex post savings to an ex ante estimate. Second was the uncertainty, even when post-program energy savings could be quantitatively confirmed, regarding their attribution to the programs themselves owing to the problem of free ridership—the recruitment of participants who would have undertaken or been likely to undertake program-recommended or subsidized actions or investments even in the absence of the program.[12] The terminology "gross" and "net" savings emerged to characterize this problem. Gross savings are those that result from program participants' actions during or following the programmatic intervention, while net savings are the proportion of these that can be attributed directly to the influence of the program itself. From the utility regulatory perspective, the importance of this distinction, among other reasons, is that regulators would want to grant incentives for DSM programs commensurate only with savings that could be demonstrated to have resulted from these programs.[13]

To illustrate these definitions, suppose that a utility estimates potential savings of 100 kWh per customer per year from a particular DSM program. The program is conducted and ex post measurements without controlling for free ridership find that participants actually saved an average of 75 kWh, while controlling for free ridership (by whatever method) results in an estimate of 50 kWh. Then the realization rate would be 0.75, while the net-to-gross ratio would be 0.66.[14]

Correctly determining gross savings is a problem of measuring energy use and changes in use in a household or firm in a way that controls for factors including weather, household composition, and other demographics, as well as economic factors that may be influencing energy demand patterns contemporaneously with the effects of a DSM program. This can involve the use of actual utility billing data as well as the use of engineering simulation or econometric models or both. Distinguishing net from gross savings in turn requires addressing the underlying self-selection problem; that is, the interaction of factors affecting voluntary program participation with those affecting changes in behavior associated with the program. In statistical models in which energy savings are estimated as a function of factors including a DSM program, the issue is that the program participation indicator is endogenous. In other areas, this problem can be solved by using randomized control groups, such as in drug treatment trials. The difficulty in DSM evaluation—as in, for example, analyses of job training programs in labor economics—is that this experimental mechanism is as a rule unavailable for analyzing effects of programs undertaken on a voluntary basis.

In addition to being the focus of a now-voluminous DSM evaluation literature, these and related issues have been the subject of scrutiny in academic studies that, although not focused on REMS programs, are nevertheless relevant. In a widely cited analysis, Joskow and Marron (1992) analyzed DSM data from mid-1991 from a national sample of programs and found consistent underreporting of costs—both administration costs incurred by the utilities in running and evaluating the programs and those incurred by participants, including their share of actual investment costs and of their transaction costs. They also found that program costs were very sensitive to estimates of energy savings, that these in turn were almost universally derived solely from ex ante engineering estimates, and that free ridership was characteristically either ignored or incompletely or incorrectly accounted for.

Eto et al. (1996) applied and extended Joskow and Marron's method to evaluate the costs and benefits of 20 commercial lighting DSM programs. They found that while there was great variation in reporting and accounting practices, all 20 were cost-effective once the problems noted above—incomplete cost inclusion, free ridership, and others—were addressed. At the same time, they strongly supported Joskow and Marron's call for standardized accounting, measurement, and reporting procedures.

More recently, Loughran and Kulick (2004) used utility data collected by the U.S. Department of Energy to estimate the electricity savings from and cost-effectiveness of DSM programs undertaken by several hundred utilities during the period 1989 to 1999. They found that the programs represented in their sample resulted in a reduction of aggregate electricity sales of between 0.3–0.4%, compared with (aggregated) utility estimates of 1.8–2.3%. They also found a much higher cost per kilowatt-hour saved than that implied in the aggregate by utility estimates. Loughran and Kulick attribute these findings to biases in utility-based estimates, the most important of which they term "selection bias." They define this term quite differently, however, than the standard definition given above: in their usage, "selection" occurs if a program participant would, in the absence of the program, undertake at any time in the future an investment or action recommended by the program. Given their additional claim that because of its embodiment in new capital, energy efficiency will be adopted by many or most households or firms over time regardless of participation or not in DSM programs, this essentially defines most or all of apparently DSM-induced efficiency gains as a selection artifact.[15, 16]

Horowitz (2004) also applies the U.S. Department of Energy's utility data to estimate the effects on U.S. commercial-sector electricity use of both standard utility DSM programs and more recent market transformation programs, which emphasize incentives to manufacturers of energy-efficient appliances and other equipment. He estimates a model that incorporates the effects not just of these programs but also of energy price changes, independent technological and economic trends, and other factors applied to data across 42 states representing 90–95% of U.S. commercial-sector electricity sales covering the period 1989 to 2001. Horowitz finds that these public programs reduced electricity intensity in this sector by nearly 8% over this period and in 2001 reduced electricity sales in this sector by an amount equivalent to 2.3% of total U.S. electricity sales in that year.

The First Protocols

While widespread improvements in the methodology and practice of DSM evaluation were underway by the end of the 1980s, the impetus for systematizing and institutionalizing these developments reached something of a critical mass by the early 1990s.[17] Under the auspices of the CPUC, a multi-stakeholder group undertook the development of a standardized protocol for the measurement and evaluation of DSM programs.[18] Their efforts culminated in the *Protocols and Procedures for the Verification of Costs, Benefits, and Shareholder Earnings from Demand-Side Management Programs* (CPUC 1998).[19] The CPUC also established an independent body, the California Demand Side Management Advisory Committee (CADMAC) and charged it with the continuing development and improvement of the *Protocols*.

As the title suggests, the *Protocols* address financial as well as measurement and evaluation aspects of DSM; the focus here is solely on the latter. First, a core set of guidelines is given that applies to all program types.[20] The guidelines state the objectives of measurement, the criteria for estimating both gross and net energy savings effects (where "gross" and "net" are as defined above), the definitions of and criteria for sampling for the purpose of measurement, and criteria for the use of billing data. The impact estimation guidelines also stipulate acceptable characteristics of load impact regression models, including model type, the type of data that they should employ, the necessity of diagnostics and test statistics, and the appropriate approach to developing individual model specifications.[21] The *Protocols* state: "Confounding effects on energy consumption should be controlled for. The use of a comparison group and the inclusion of social, political, and economic changes, are acceptable methods (*12*)."

The guidelines also explicitly allow for statistical models that have "observed decisions of customers to participate in DSM programs and to install efficient equipment as the dependent variables. The purpose of these models is to control for free-ridership or to derive a net-to-gross savings adjustment. The models may also be used to estimate an adjustment factor to control for self-selection bias (*13*)."

These general guidelines are complemented by a set of guidelines specific to 11 program types in addition to energy management services.[22] For the latter, these include designated units of measurement and criteria for data quality and specific measurement issues for both participants and comparison groups. In addition, the *Protocols* require long-run follow-up or "persistence" studies for several categories of programs, extending as far as nine years after the program year. Such studies, however, were not required for informational programs, including REMS programs.

REMS Program Evaluation in Practice

Following are three examples of evaluations of REMS programs for single-family dwellings conducted in the 1990s in the several years following the introduction of the *Protocols*. These programs are based on energy audits of individual households. Through the use of billing data, in-home inspections by utility staff, and surveys, the utility estimates for each participating household potential energy savings both from purchases of specific pieces of equipment, such as appliances or thermal shell improvements, and from changing energy usage practices, such as thermostat management, temperature settings for refrigerators, and so forth. These estimates are then provided at no direct cost to each household, but no payment or other financial incentive is provided. The following program-specific reviews draw on documentation available from the California Measurement Advisory Council, the successor organization to the aforementioned CADMAC, as well as on data of the CEC. One of these programs focused on electricity, one on natural gas, and one on both. Each used billing data and a statistical model to estimate gross and net savings and each included a comparison group.

The requirements in effect at the time of these evaluations did not stipulate whether or how program-savings estimates should be reported in terms of what would seem to be the most natural index, some form of the percentage electricity or gas consumption reduction relative to a measurement of prior or baseline consumption. Indeed, the definitions of gross and net savings in the *Protocols* do not define a specific benchmark for this kind of comparison. Naively, it might seem that a pre-to-post program decline in absolute levels of energy consumption should be the criterion but clearly this is too stringent: unusually hot or cold weather during the program year, for example, or a change in household composition might result in an increase in absolute consumption even as effective conservation and efficiency actions are being undertaken. When statistical models are used, as in the cases below, both net and gross savings implicitly are defined as the values of estimated coefficients of variables indicating program occurrence and participation.

Nevertheless, estimated program-induced savings as a percentage of some measurement of baseline consumption are intuitively the most natural way of gauging program effects. The documentation does not in all cases, however, either report results in this form or provide sufficient information for them to be extracted. In only one of the following three cases did the evaluator provide an estimate of savings in percentage terms. In the other two cases, estimates for this chapter, which are summarized in Table 8-2, were obtained indirectly using CEC data; the details are provided below.

Pacific Gas & Electric

Pacific Gas & Electric's (PG&E) 1997 Residential Energy Management Services Program served both single-family and multi-family dwellings; only the single-family case is discussed here.[23]

The program provided free-of-charge household-specific energy-efficiency and energy-use information pertaining to heating and cooling-related appliances, systems, and building envelope characteristics based on a customer's response to a mail survey or phone survey, or participation in an on-site audit. It also included a service to directly answer customers' energy-efficiency questions. Approximately 100,000 responses to the mail survey, 9,000 on-site audits, and 1,000 phone surveys were reported. Following the program, the evaluation disseminated a follow-up survey to participants and a randomly chosen sample of nonparticipants from the utility's customer database that gathered information on household (demographic) characteristics, stock of energy-using equipment, and energy-related practices and asked about energy conservation or efficiency actions in the three years prior to the time of the audit recommendations. Following cleaning of the data, the final sample of completed surveys contained approximately 800 nonparticipants and 1,800 participants. These data were combined with weather and billing data to yield a monthly record for each household, including both electricity and gas consumption, covering the period from January 1996 to September 1998, to carry out the evaluation.

Savings estimates were obtained using a linear regression model with monthly electricity or gas as the dependent variable and weather, demographic, equipment, energy-saving behaviors, and a binary variable for program participation. In addition to experimentation with specification through different variable interactions, the models were estimated in three ways: total consumption-by-month through the entire sample; total annual consumption with pre-program consumption being average daily consumption during calendar year 1996 and post-program consumption average daily consumption from January through September 1998; and a model identical to the latter except with the pre-program period shortened to January–September 1996. The evaluators selected the annual model with nine months each of pre- and post-program consumption and all variables except participation variables interacted with heating or cooling degree days. The evaluators report that the other two were rejected due to instability, which they suggest was due to "weather issues:" "the effect of including heating and cooling degree days and consumption from October through December 1996 but from the corresponding dates of 1998 may have been to increase the variability of the data beyond the power of the model to resolve...[possibly] exacerbated by the fact that, due to El Niño and La Niña, the weather during the three years under analysis varied more than usual." All coefficient estimates for the final model were significant at the 95% level. There was no reported treatment of other statistical issues such as testing for and correcting heteroskedasticity or autocorrelation.

The evaluation estimates net electricity savings of 240 kWh per participating household per year, with a 90% confidence interval of 58-to-421 kWh per year. Peak demand savings per participating household per year were estimated as 0.043 kilowatts (kW), with no confidence interval reported. However, both participants' and nonparticipants' consumption and demand were estimated to have increased in absolute terms from pre- to post-measurements, illustrating the point above regarding the definition of "savings" in this context. Ex ante program-savings estimates were generated by the utility's internal planning and forecasting apparatus. The realization rate for kWh was 2.65, with a net-to-gross ratio of 0.63, and for kW 2.66 and 0.63. That is, the ex post estimates of net program savings were more than twice as high as the ex ante estimates, while in comparison with the nonparticipant group, 63% of participants' estimated savings were attributed to the program itself. Put differently, consumption and demand of both participants and nonparticipants grew from pre-to-post but the increase was slower for participants and this difference was attributed to the program.

Net gas savings were estimated as 10 therms per participating household per year, with a 90% confidence interval of a 0.28 increase in consumption to a 20.6 therm savings. The realization rate for natural gas was 1.01. As with electricity consumption and peak demand, both participants' and nonparticipants' gas consumption increased in absolute terms on a pre-to-post basis. The gross change estimate for natural gas, however, was reported to be positive; that is, the program was statistically estimated to have increased consumption prior to comparison with nonparticipants. Aside from the fact that the confidence interval for net savings includes zero, this makes the notion of "net savings" for gas rather confusing in this case. What is meant literally is that the estimated coefficient on the participation dummy is positive when the model is estimated only on participants and negative when estimated on both participants and nonparticipants. One possible interpretation is that participants' gas consumption increased as a result of the program but not as much as nonparticipants' increased in the absence of the program. In any case, however, designating this outcome as a "net savings" is questionable.

Notwithstanding this anomaly in the natural gas results, the evaluation document reports neither net savings for electricity or gas as a percentage of pre-program or baseline consumption nor provides sufficient information for estimates of this type to be calculated. Therefore, the evaluation data were combined with CEC data to obtain approximate percentage estimates as follows.[24] For electricity consumption, CEC estimates for electricity consumption across end-uses for single-family dwellings in the PG&E service territory by end-use for 1990 and 2000 were aggregated and the result interpolated to 1997. This was combined with an estimate of the number of such dwellings in the service territory in 1997 to obtain an estimate of annual consumption by household and in turn the mean REMS net savings per household as a percentage of this quantity of 2.94%. Because no estimates were available by dwelling type for peak demand, the evaluation net-savings estimates were estimated as a percentage of mean residential peak demand in 1997 in the PG&E service territory for all dwelling types—2.8%. These results are summarized in Table 8-2.

TABLE 8-2. Outcomes of Residential Energy Management Services Programs in California

	Energy type	Number of participating households	Realization rate	Net-to-gross ratio	Estimated savings per household per year (90% confidence interval)	Estimated savings as percentage of annual household consumption or peak demand
Pacific Gas & Electric, 1997	Electricity	109,424	2.65	0.63	240 kWh (58, 421)	2.9%
	Peak demand		2.66	0.63	43 W	2.8%
	Natural gas		1.01	n/a	10 therms (−0.28, 21)	2.1%
SoCal Gas, 1997	Natural gas	21,133	0.65	n/a	29 therms (18, 40)	4.56%
Southern California Edison, 1995	Electricity	10,000	1.1	0.72	343 kWh (237, 448)	4.5%
	Peak demand		1.8	0.72	74 W (51, 97)	4.1%

Source: Numbers of participants, realization rates, net-to-gross ratios, savings in physical units, and percentage savings in SoCal Gas program are taken from RER (1997), Goett (1999), and Hagler-Bailly (1999). Percentage savings estimates for PG&E and SCE programs are author's calculations; see text.

Southern California Gas

Southern California Gas' (SoCal Gas) 1997 "Home Energy Fitness" program provided informational audits at no charge to customers.[25] The utility estimated a given household's gas demand by end-use using past consumption data from billing records and a detailed questionnaire about equipment and practices, such as settings on equipment, typical times of use, and so forth. Participating households then received, along with the disaggregated consumption estimates, a set of recommendations for energy-saving, such as the purchase of high-efficiency units and thermal insulation, reducing hot water usage, wrapping water heaters, and turning off equipment when not in use, as well as an estimate of the cost savings that would accrue from implementing the recommendations.

The process began with a targeted mailing to 200,000 customers in single-family dwelling units (detached homes, condominia, and townhouses), with recipients initially selected by the criterion of having had at least 10 years of service from the utility. About 23,000 customers responded and received an audit. In preparation for the evaluation, SoCal Gas also provided to the evaluator a randomly drawn list of customers consistent in length of service and dwelling type from which to draw a comparison group. The participants were stratified by length-of-service, weather zone, and average daily consumption. For a follow-up survey, a sample of participants was drawn proportionately from the overall set of participants and a corresponding sample of nonparticipants drawn to match the stratification. A telephone survey was administered with questions on demographic characteristics, household income, dwelling characteristics, and a series of energy-related questions focused on investments in gas-using equipment and changes in utilization practices of such equipment during the previous two years. The final sample contained about 560 participants and 500 nonparticipants. On demographic questions, equipment ownership and changes, and conservation practices, the differences between the two groups were reported as statistically insignificant or moderate, while dwelling sizes and ages were significantly different. Finally, 27 months of billing and weather data from September 1996 through November 1998, covering about one year before and one year after the audits, were combined with the survey data. After some respondents were dropped due to missing data problems, about 550 participants and 500 nonparticipants were contained in the final samples used in the evaluation.

The evaluator applied a linear regression model with daily gas use a function of binary variables representing the presence or absence of specific equipment types, the weather zone in which the dwelling was located, heating and cooling degree days, and three binary participation and effect variables, one representing program participation, one distinguishing pre-and-post 1998 (i.e., before and after the energy audits), and a third that took the value 1 only for program participants after receipt of the audit recommendations. The evaluator interpreted the second of these as measuring the "combined effect of economic and environmental trends, as well as impacts of any 'naturally' occurring conservation actions (4)." When entered with the 1998 dummy, this post-audit participation dummy was interpreted as measuring the net impact of the program after controlling for naturally occurring conservation and trends. Here "naturally occurring" apparently is intended to mean "actions that would have been taken in the absence of the program," encompassing free-rider actions (1).

Both autocorrelation and heteroskedasticity were detected and corrected. The evaluator experimented with 12 model specifications and selected one preferred specification on the basis of explanatory power, statistical significance, and the plausibility of both the parameter estimates and the predicted effects. The net savings were calculated using the parameter estimates from the preferred model as a function of the participation and effect variables above combined with the weather variables (and reported along with confidence intervals). The estimated net savings were 29 therms per participant per year, with a 90% confidence interval of 18 to 40 therms. An evaluation of the "Home Energy Fitness" program for 1994 had found 44 therms per year per participant in net savings; this was used as the ex ante savings estimate, giving a realization rate of 0.65. For reasons that are not reported, neither a gross savings estimate nor a net-to-gross ratio was given.

In this evaluation, net savings were estimated by the evaluator as a 4.56% reduction (see Table 8-2). Here again, the definition comes into play. This reduction was as a percentage of participating households' mean consumption in the year prior to the program. The pre-to-post change in consumption for participants, however, was a 24% increase. Thus, as in the previous case, "savings" constitute a moderation of an increase in absolute consumption. It also is worth noting that this increase is extremely high; the evaluation suggests the effects of the "El Niño" weather anomaly.

Southern California Edison

Southern California Edison's (SCE) 1995 "In-Home Audit Program" provided energy-conservation recommendations and information about high-efficiency appliances to interested residential customers free-of-charge.[26] Both telephone and on-site versions of the program were offered; the documentation describes the two as being "essentially identical." The program recommendations concerned both behavioral changes or practices and equipment changes. The latter information regarded the characteristics and availability of high-efficiency appliances only; the program did not make specific purchase suggestions for equipment.

The initial sample for the evaluation contained all 10,000 1995 program participants and 50,000 nonparticipants. Screening and cleaning of the data (e.g., to eliminate participants who had either ended service or who had not had service in the prior year) resulted in a sample size of approximately 7,000 participants and 33,000 nonparticipants. Following sample stratification and other processing, sub-samples were drawn to administer the post-program evaluation survey. The final sample contained 1,200 participants and nonparticipants each. SCE provided consumption data for every subject for the period January 1993 through September 1996, which was in turn combined with weather data to complete the dataset. Participant and nonparticipant samples for the evaluation were compared on economic and demographic characteristics, as well as on appliance holdings and end-use actions but no statistics were given beyond mean values. Informally, there was a fair degree of non-comparability on a number of dimensions. For example, household income was about 21% higher for participants than nonparticipants, almost 50% more participants than nonparticipants had central air conditioning, and nearly three times as many participants had room air conditioners.

The evaluators developed a statistical model in stages. They first formulated essentially a difference-in-differences model in which the dependent variable was the 12-month change in household consumption in kilowatt hours and the regressors were indicator variables for the presence of specific pieces of equipment, variables for changes in equipment, household characteristics, and weather, as well as a dummy variable for program participation, along with various interaction terms. To address the selection problem, the model was augmented with a Double Mills Ratio specification in which a inverse Mills ratio was derived from an equation in which program participation was modeled as a function of site characteristics, observable demographic and economic characteristics of the household, and an indicator of weather conditions and then entered into the previous model both directly (i.e., as a stand-alone variable) and interacted with the participation indicator.[27] The model was then estimated twice, once with and once without the Double Mills augmentation. The basic version of the model yielded an estimate of 55-kWh savings per month per participant; the augmented model yielded a higher estimate of 60 kWh per month per participant. The evaluators observed that both estimates were implausibly high for this program type and conjectured that the self-selection bias might not have been resolved or that neither model may have accounted correctly for weather conditions (or possibly both).

The final model was a "composite end-use" approach that integrated into the basic regression model terms involving specific behavioral actions and other details, as well as the Double Mills Ratio. It disaggregated estimated savings both by end-use categories and by measures and practices (e.g., equipment and behavioral changes). End-use savings were estimated using engineering models and then interacted with "participation" dummies indicating specific actions, such as an equipment purchase or a change in the utilization of an appliance.

The realization rate for energy was 1.1 and for demand was 1.8. The net-to-gross ratio was 0.72 for both. Estimated net-electricity savings per household per year were 343 kWh with 90% confidence interval 237 to 448 kWh. Estimated peak-demand reduction per household was 74 W with 90% confidence interval 51 to 97 W. As with the PG&E program, savings in percentage terms were not reported and therefore were estimated for this chapter using the same data and procedure as in that case. The result was an estimated 4.5% reduction in electricity consumption per household per year and 4.1% reduction in peak demand. The document did not provide pre- and post-program usage information that would indicate whether these savings constituted an absolute decline or a slowing of energy and demand growth (see Table 8-2).

Interestingly, it was determined that changes in practices—the manner in which the existing equipment stock was used—rather than changes in the equipment itself accounted for all program-induced savings and that more than 73% of these were refrigeration actions, including disconnecting second refrigerators, checking condenser coils, keeping units full, and lowering thermostat settings.

Finally, the evaluation included an assessment of whether and how the program overcame the "information market barrier"—whether households simply had been unaware of the available energy-savings actions that were described to them by the program. For each of the 77 energy-efficiency action items covered by the program, respondents were asked first whether they previously had been aware that the item could save energy if employed and second whether prior to the audit they had realized the importance of the action. Overall, it was found that customers were not subject to this information barrier. Most respondents already were aware of the 77 action items, although in the case of 13 of these measures, at least 50% of the participants had not been aware of them. While it might be suggested that this result indicates some sort of superfluity of the program intervention, the evaluators conclude that "there is some support for the idea that, while vaguely aware of the energy benefits of the recommended actions, customers do not always act on this knowledge until it is specifically suggested by an Edison expert."

In the view implied by this conclusion, an alternative interpretation is that the binary model of customers "having" or "not having" information on energy efficiency and conservation is an inadequate description of how customers' knowledge and decision processes actually operate and that a more nuanced understanding of how the program actually affected customer actions would be appropriate.

Discussion

By way of background, it is important to note that the selection problem in DSM evaluation has been the subject of technical research and experimentation for several decades, concurrent with the policy and institutional developments described above, and that these efforts are continuing. The ideal solution to this problem—randomized assortment of program candidates into participating and non-participating groups—is, as observed previously, difficult or impossible by definition in the context of voluntary programs. The practical alternative, as sanctioned in the *Protocols*, is to construct a comparison group of nonparticipants that is matched on observable characteristics, ideally along demographic dimensions such as age, income, and household composition, as well as on both technical dimensions, such as dwelling characteristics and ex ante appliance stocks and weather influences. Such matching can be challenging in practice, and in the absence of formal techniques it may be difficult to determine the extent to which it corrects for unobservable characteristics of program participants that may be the source of self-selection bias. Propensity-score matching methods for non-randomized treatment and comparison groups developed in labor economics would appear well-suited to DSM evaluation, although they have yet to be applied.[28]

The models applied in the PG&E and SoCal Gas evaluations are variations of what is known as "conditional demand analysis" in the DSM evaluation literature, in that energy consumption is estimated in part as a function of equipment holdings. While including both pre- and post-program consumption information and some other dynamic elements, however, they are not strictly speaking difference-in-differences models, which under certain assumptions described below can yield unbiased estimates of program effects. Rather, the means to address the selection problem is in both these cases the use of the comparison or nonparticipant group.

In the case of PG&E, the comparison group was drawn from a simple random sample of nonparticipants from the utility's customer database and screened solely for completeness of records and anomalies without any attempt at matching on characteristics. Indeed, the evaluation reports that nonparticipant electricity consumption only was roughly two-thirds of participant consumption, both before and after the program. Methodologically, therefore, self-selection was addressed in a very limited fashion in this evaluation.

By contrast, in the SoCal Gas evaluation fairly close attention was paid to matching participants and nonparticipants on observable characteristics and there was a reasonable correspondence on a number of relevant dimensions, though not all. Thus, in this case, also given the more careful specification of participation variables in the model, self-selection was more fully addressed methodologically than in the previous example. In addition, one could have greater confidence in these results given that standard estimation problems were recognized and dealt with.

As noted above, interpretation of both sets of results depends considerably on how "net savings" are defined and how this definition is interpreted. In these evaluations, the definition is, strictly speaking, purely a statistical one and "savings" are registered even as consumption is increasing. While this is not unreasonable, it would be warranted that it be stated explicitly. At the same time, the reporting of net savings in gas consumption in the PG&E study, where the program actually is estimated to have increased consumption relative to a baseline, strains credulity.

Although the SCE evaluation did not emphasize the matching of the comparison to the participant group, it nonetheless exhibited considerably greater sophistication in modeling. It is curious, however, that the evaluators did not discuss the characteristics of their initial model that might have reduced the selection bias problem. It has long been recognized that difference-in-differences models originally developed in labor economics could be adapted to DSM program evaluation when pre-and-post program data were available to correct for self-selection bias, even in the absence of a randomized control group (Heckman and Robb 1982; Ozog and Violette 1990). The key assumption is that the unobserved factor or factors determining self-selection are stable over the evaluation period. In the DSM context, for example, an underlying difference in disposition to invest in efficiency or conservation between participants and nonparticipants would be corrected for provided it did not change before, during, or after the program. This would appear plausible, although not guaranteed; for example, this propensity might change as a result of weather events during the evaluation period that might "interact" with an underlying propensity toward conservation and efficiency on the part of participants (although this might be addressed through the manner in which weather variables entered the model). The Double Mills Ratio has been shown to correct for self-selection when the distribution of net savings over customers is approximately normal (Goldberg and Train 1996). The validity of this assumption in this particular program cannot be tested with the available information. Nevertheless, from a methodological standpoint, this evaluation addressed the self-selection problem in the most rigorous manner of the three evaluations.

The use of the term "methodological" in these remarks is deliberate. While the use of less or more appropriate techniques for correcting potential selection bias can be examined, strictly speaking the actual underlying magnitude of the problem in specific cases, and therefore the utility of more sophisticated methods or their effectiveness in correcting the problem, cannot be known without a different kind of investigation. Further testing of specific cases using methods to estimate the magnitude of self-selection as well as to correct for it—whether on simulated datasets as in Goldberg and Train (1996) or with actual program data as in Kandel (1999b)—would be a valuable contribution to the DSM evaluation literature.

Overall, the magnitudes of the estimated savings from the three programs are consistent with the generally expected magnitude—up to the neighborhood of 5%. For the reasons stated above, the most confidence reasonably could be assigned to the SCE savings estimate and the least to the PG&E. It is

interesting that for electricity consumption and demand, the program sub-
jected to the most careful scrutiny (SCE), including the treatment of
self-selection, is estimated to have higher savings than the one in which the
evaluation was weak (PG&E). The plausibility of the SCE estimates is, in any
case, supported by the finding that the preponderance of savings came from
changes in the management of household refrigerators, particularly the discon-
nection of second units, which are likely to be older and therefore consume
more energy than the primary units.

While this small a number of examples only can be considered illustrative,
these three evaluations suggest that the development and implementation of
the *Protocols* were reflected in substantial improvements in DSM evaluation relative to the earlier era but that it did not solve in practice all the problems that
had been raised regarding the measurement of savings. Shortcomings are not
surprising given that the guidelines under which they were conducted were
new and a certain amount of trial-and-error would be expected. Practical quality control was recognized in the *Protocols* as an issue that needed to be
addressed explicitly and institutionally. [29] Further remarks related to these
points are offered in the following subsection.

Additional Findings on Controlling for Selection Bias

The *Protocols* had been in effect for a relatively short time when California dramatically shifted its approach to electricity regulation and policy.[30] Assembly
Bill 1890, passed by the state legislature in 1996, called for the restructuring of
the state's electric power system. The spirit and the letter of this legislation
were to substantially reduce the state's regulatory footprint in the generation,
transmission, and sale of electricity in California. Utility DSM in its traditional
form was no longer sanctioned, but for policy and political reasons, a public
role in the electricity system was maintained through the introduction of a
public goods charge, an electricity tax to fund research and development related
to the public, including environmental, consequences of electricity generation
and a transformed version of DSM.

Under the law, the traditional justifications for utility DSM officially ceased
to be recognized, in a manner of speaking, but a role was acknowledged for
public policy in promoting energy efficiency through the "market transformation" approach—primarily by using public-goods-charge revenues to make
incentive payments to manufacturers to lower the cost of the efficient equipment they sold in the market. The *Protocols* were no longer in force after the
1998 program period, although persistence studies pursuant to first-year evaluations conducted while the *Protocols* were in force continued to be carried
out.

The electricity supply crisis of 2000–2001, however, began a shift back
toward a public role in the electricity system in general and the market for

energy efficiency in particular, not just for reducing energy consumption but, especially during the height of the crisis, for its demand or load-management benefits as well. The following years have seen not just a return to, but a dramatic expansion of, public promotion of energy efficiency in California, with both more funds from the public-goods charge and, as of 2004, a re-authorization of the investor-owned utilities to augment energy-efficiency programs with rate-based funding.[31] The current era of energy policy in California is again to a large extent grounded in the priorities first articulated in the Warren-Alquist Act more than 30 years ago. A significant new element, however, is an aggressive effort to develop and implement a comprehensive strategy for reducing statewide carbon dioxide emissions as a response to the risks of global climate change.[32]

Concurrent with these policy initiatives has been a return of attention and resources to program evaluation policy, methodology, and implementation. In 2002, the CPUC adopted on an interim basis the *Energy Efficiency Policy Manual* based on the U.S. Department of Energy's International Measurement and Verification Protocol. A major planning process was instituted to create a new evaluation approach for California's public energy efficiency programs, culminating in the *California Evaluation Framework*, issued in 2004 (TecMarket Works Framework Team 2004). This "provides a consistent, systematized, cyclic approach for planning and conducting evaluations of California's energy efficiency and resource acquisition programs." The framework addresses theoretical, empirical, practical, and institutional issues across a range of topics related to evaluation.[33] It has been followed by a new set of evaluation protocols specifying detailed criteria, guidelines, and methods for evaluation in practice (TecMarket Works Framework Team 2006). These institutional developments are being matched by resources for practical application; of $2.5 billion authorized for expanded energy efficiency deployment in 2006–2008, a full 8% is earmarked for evaluation, measurement, and verification.

The new framework addresses a number of aspects of the estimation of net-to-gross savings ratios and the treatment of selection bias for DSM in general. It is worth quoting at length:

> ...we need [in evaluating energy efficiency programs] to answer the question of what would participants (and nonparticipants) have done in the absence of the program that could affect their energy use level. This sounds like a simple question. But 15-20 years experience has proven it to be a very difficult question to answer with any assurance as to the (unbiased) accuracy or precision of the answer.
>
> These doubts in the quality of these estimates are not for lack of trying. The increasing study and sophistication of evaluation efforts to improve upon this capability throughout the 1990s was far more extensive in California than anywhere else. Thorough analysis by expert

econometricians has repeatedly found problems in each new generation of "solutions." Nevertheless, this section [on net-to-gross analysis and the treatment of selection bias] will summarize and reference several recent methods as possible choices.[34]

Because survey methods (e.g., using self-reported data on participant and non-participant actions related to programmatic interventions as a means of controlling selection bias) are widely used for their practicality and affordability, these methods are reviewed in the framework and guidelines recommended for their appropriate application. The econometric analysis of selection bias also is reviewed at length, including the history of the evolution of this methodology. The authors describe the use of the inverse and double inverse Mills Ratio, discussed in the previous section. They also discuss the more recent development of the so-called instrumented decomposition approach due to Kandel (1999a), which, although developed for appliance-rebate rather than information programs, appears to represent the current state-of-the-art in the econometric analysis of net savings controlling for selection. It is a two-stage approach extending previous work by Train et al. (1994) based on decomposing post-program energy bill changes into savings from the rebated appliance and savings from trends in underlying economic factors, weather, and other factors. In the first stage, a nested multinomial logit model is used to predict households' underlying propensity to purchase the appliance; the results are used to create instruments for a second regression incorporating interaction terms for free ridership and participation. Application of this method to an evaporative cooling rebate program showed lower-than-expected rates of free ridership—around 20%—and therefore greater load reductions than previously estimated (Kandel 1999b).

Conclusion

The critical literature on utility DSM has tended to emphasize inadequate ex post measurement and the treatment of self-selection and often concluded that these programs as a rule provide lower energy savings at greater cost than standard estimates would suggest. Developing and applying methods to ensure verified savings from energy-efficiency programs, however, has been the focus of a significant, publicly funded effort in California for several decades. The three evaluation examples reviewed in this chapter illustrate both the practical aspects of dealing with measurement and selection problems and the fact that they can be and have been addressed in that state. By virtue of the methods applied and the empirical results, it can be concluded that at least two of the three evaluated residential DSM programs resulted in energy savings on the order of several percent, a magnitude consistent with previous findings on successful programs of this type.

There nevertheless remain technical challenges to ensuring that selection bias in DSM evaluation is corrected routinely, and many years of research have demonstrated that there is no silver bullet for this problem. A valuable research priority would be further studies to estimate the extent of selection bias in specific cases and the value of more sophisticated econometric methods for overcoming it. Such research could be valuable for regulators in gauging how to balance the further allocation of resources to reducing the selection bias problem as opposed to other evaluation issues that also are important for reliably measuring program savings.

References

Blumstein, Carl, Betsy Krieg, Lee Schipper, and Carl York. 1980. Overcoming Social and Institutional Barriers to Energy Conservation. *Energy* 5: 355–371.

Brown, Marilyn A., and Philip E. Mihlmester. 1994. *Summary of California DSM Impact Evaluation Studies.* Report ORNL/CON-403. Oak Ridge, TN: Oak Ridge National Laboratory.

CalEPA (California Environmental Protection Agency). 2006. *Climate Action Team Report to Governor Schwarzenegger and the Legislature.* Sacramento, CA: CalEPA.

Carnahan, Walter, et al. 1975. *Efficient Use of Energy: A Physics Perspective: A Report of the Summer Study on Technical Aspects of Efficient Energy Utilization, July 1974.* College Park, MD: American Physical Society.

CEC (California Energy Commission). 2003a. *California Energy Demand 2003-2013.* CEC Staff Report #100-03-002. Sacramento, CA: CEC.

———. 2003b Mike Messenger, principal author. *Proposed Energy Savings Goals for Energy Efficiency Programs in California.* CEC #100-03-021. Sacramento, CA: CEC.

CPUC (California Public Utilities Commission). 1998. *Protocols and Procedures for the Verification of Costs, Benefits, and Shareholder Earnings from Demand-Side Management Programs.* San Francisco, CA: CPUC.

Dehejia, Rajeev H., and Sadek Wahba. 2002. Propensity Score-Matching Methods for Nonexperimental Causal Studies. *The Review of Economics and Statistics* 84(1): 151–161.

Eto, Joseph. 1996. The Past, Present, and Future of U.S. Utility Demand-Side Management Programs. Report LBNL-39931/UC-1322. Berkeley, CA: Lawrence Berkeley National Laboratory.

Eto, Joseph, Leslie Shown, Richard Sonnenblick, and Christopher Payne. 1996. The Total Cost and Measured Performance of Utility-Sponsored Energy Efficiency Programs. *The Energy Journal* 17(1): 31–52.

Fels, Margaret, and Kenneth M. Keating. 1993. Measurement of Energy Savings from Demand-Side Management Programs in US Electric Utilities. In *Annual Review of Energy and the Environment* (vol. 18), edited by Robert H. Socolow, Dennis Anderson, and John Harte. Palo Alto, CA: Annual Reviews, 57–88.

Fischer, Carolyn. 2005. On the Importance of the Supply Side in Demand-side Management. *Energy Economics* 27: 165–180.

Goett, Andrew A. 1999. *1997 Residential Energy Management Services First Year Load Impact Evaluation (Home Energy Fitness Program): Study ID No. 715.* CA: AAG & Associates.

Goldberg, Miriam, and Kenneth Train. 1996. *Net Savings Estimation: An Analysis of Regression and Discrete Choice Approaches.* Madison, WI: Xenergy, Inc.

Hagler-Bailly Consulting. 1999. *Impact Evaluation of Pacific Gas & Electric Company's 1997 Residential Energy Management Services Program (Single Family): PG&E Study ID #397.*

Heckman, James, and Richard Robb. 1982. Alternative Methods for Evaluating the Impact of Interventions. In *Longitudinal Labor Market Studies: Theory, Methods, and Empirical Results,* edited by J. Heckman and B. Singer. NY: Academic Press.

Horowitz, Marvin J. 2004. Electricity Intensity in the Commercial Sector: Market and Public Program Effects. *The Energy Journal* 25(2): 1–23.

Huntington, Hillard, Lee Schipper, and Alan H. Sanstad (eds.). 1994. Markets for Energy Efficiency. *Energy Policy* 22(10): October.

Jaffe, Adam B., and Robert N. Stavins. 1994. The Energy-Efficiency Gap: What Does It Mean? *Energy Policy* 22(10): 804–810.

Joskow, Paul L., and Donald B. Marron. 1992. What Does a Negawatt Really Cost? Evidence from Utility Conservation Programs. *The Energy Journal* 13(4): 41–74.

Kandel, Adrienne. 1999a. Instrumented Decomposition: A Two-Stage Method for Estimating Net Savings. *Proceedings of the 1999 International Energy Program Evaluation Conference,* Denver, CO.

———. 1999b. Evaporative Cooler Rebate Program Cuts Load Significantly, and May Overcome Class Barrier. *Proceedings of the 1999 International Energy Program Evaluation Conference,* Denver, CO.

Loughran, David S., and Jonathan Kulick. 2004. Demand-Side Management and Energy Efficiency in the United States. *The Energy Journal* 25(1): 19–43.

Nadel, Steven. 1992. Utility Demand-Side Management Experience and Potential: A Critical Review. *Annual Review of Energy and the Environment* 17: 507–535.

NARUC (National Association of Regulatory Commissions). Conservation Committee. 1989. *Resolution in Support of Incentives for Electric Utilities' Least-cost Planning.* Washington, DC: NARUC.

Ozog, Michael, and Daniel Violette. 1990. Using Billing Data to Estimate Energy Savings: Specifications of Energy Savings Models, Self-Selection, and Free-Riders. In *Proceedings of the ACEEE 1990 Summer Study on Energy Efficiency in Buildings Panel 6: Program Evaluation,* Washington, DC.

RER (Regional Economic Research, Inc.). 1997. *1995 In-Home Audit Evaluation: Study ID No. 528 (A).*

Rosenfeld, Arthur H. 1999. The Art of Energy Efficiency: Protecting the Environment with Better Technology. *Annual Review of Energy and the Environment* 24: 33–82.

Sanstad, Alan H., W. Michael Hanemann, and Maximillian Auffhammer. 2006. The Role of Energy Efficiency. In *The Potential Economic Costs and Benefits of Climate Change Policy in California,* edited by W. Michael Hanemann and Alexander E. Farrell. San Francisco: The Energy Foundation.

TecMarket Works Framework Team. 2004. *The California Evaluation Framework.* Oregon, WI: TechMarket Works.

———. 2006. *California Energy Efficiency Evaluation Protocols: Technical, Methodological and Reporting Requirements of Evaluation Professionals.* Oregon, WI: TechMarket Works.

Train, Kenneth, S. Buller, B. Mast, K. Parikh, and E. Paquette. 1994. Estimation of Net Savings for Rebate Programs: A Three-Option Nested Logit Approach. *Proceedings of the ACEEE 1994 Summer Study on Energy Efficiency in Buildings,* Asilomar, CA.

Vine, Edward, Chang-Ho Rhee, and Keun-Dae Lee. 2004. Electric Utility Restructuring and Its Impact on Energy Efficiency and Measurement and Evaluation: Two Unfinished Stories (California and South Korea). *Proceedings of the 2004 Summer Study on Energy Efficiency in Buildings,* Pacific Grove, CA.

Walker, James A., Theador N. Rauh, and Karen Griffin. 1985. A Review of the Residential Conservation Service Program. *Annual Review of Energy* 10: 285–315.

Notes

1. I would like to thank Billy Pizer and Dick Morgenstern for soliciting my participation in this project, helping develop the topic for this chapter, and for invaluable review comments and technical advice. I would also like to thank Ed Vine, without whose very generous guidance in understanding the history of DSM evaluation and accessing the literature I could not have written this chapter, and an anonymous RFF referee for comments. Any errors of omission or commission are solely those of the author, as are the opinions expressed herein, which should not be ascribed to the Berkeley Lab, the U.S. Department of Energy, or any other agency, institution, or individual.

2. Rosenfeld (1999) is a fascinating history of end-use energy efficiency analysis and policy by one of the principle developers of both.

3. This passage draws upon Eto (1996).

4. In this context, a "hurdle rate" is the internal rate-of-return required by consumers or firms for efficiency investments, imputed ex post from data on actual investments of this type.

5. A range of views on these issues was presented in the October 1994 issue of the journal *Energy Policy* (Huntington et al. 1994). Jaffe and Stavins (1994) therein is a widely cited discussion. Sanstad et al. (2006) provide a more recent analysis in the context of California's current long-term energy and environmental goals.

6. It is also worth noting that recent work has shown that the economic properties of minimum energy-efficiency performance standards depend upon market structure and that under certain conditions such standards can be welfare enhancing (Fischer 2005).

7. At least until restructuring of the electric power system in California in the mid-1990s, this debate was closely intertwined with, although not identical to, a concurrent debate over the validity of utility-based DSM programs from the standpoint of regulatory economics. The latter had to do with the justification of such programs as part

of least-cost planning of electricity supply, in particular the economic accounting under which end-use efficiency investments by investor-owned, publicly-regulated utilities were justified as substitutes for investments in new generation capacity. This debate also arose in part from disagreements over the existence of economic inefficiencies corresponding to persistence of technical inefficiencies with regard to energy efficiency that persisted in the market even in the face of apparently cost-effective alternatives.

8. These developments are discussed on pages 158–160.

9. The CEC does not report natural gas savings in a comparable form.

10. In the case of the historical CEC data, Messenger notes the importance of these changes in estimating the overall costs of utility DSM in California in terms of savings per dollar (CEC 2003b).

11. This section draws on Vine et al. (2004).

12. This use of the term "free rider" is in contrast to its conventional use in microeconomics to refer to economic agents who benefit from the provision of a public good without contributing to its production. I thank Billy Pizer for pointing this out.

13. Ozog and Violette (1990) point out that the selection problem also can lead to underestimating program savings in cases in which customers who opt not to participate may be those who have already undertaken the efficiency and conservation actions, so that participants will have higher initial savings potentials.

14. Although the distinction is not always made clear in the literature, this refers to "gross realization rate." The "net realization rate" would be 0.5.

15. Loughran and Kulick's framing of the problem also fails to take into account that the policy-relevant outcomes of DSM are not wholly time invariant: When, as in California, these investments are undertaken in part for load management purposes, their timing is critical, and "asymptotic accounting" does not correctly measure these benefits.

16. Both Joskow and Marron (1992) and Loughran and Kulick (2004) also point out the need for estimating program savings over time; that is, beyond the year in which the program is conducted. As noted subsequently in the text, such "persistence" studies came to be required by California in the mid-1990s for several categories of DSM, although not informational programs, and in some cases are still being conducted for programs carried out nearly a decade ago.

17. Brown and Mihlmester (1994) provide a comprehensive review of the state-of-the-art in DSM evaluation in California at that time.

18. The participants were the Pacific Gas & Electric Company, the San Diego Gas & Electric Company, Southern California Edison, Southern California Gas, the California Energy Commission, the CPUC Office of Ratepayer Advocates, and the Natural Resources Defense Council.

19. Adopted in CPUC Decision 93-05-063, later revised and amended.

20. CPUC (1998), Table 5, "Protocols for the General Approach to Load Impact Measurement."

21. Numerical modeling for DSM evaluation is the focus of its own subset of the eval-

uation literature; see Fels and Keating (1993).

22. These include, for example, residential weatherization, residential appliance efficiency incentives, and efficiency incentives for each of the commercial, industrial, and agricultural sectors.

23. The evaluation report is Hagler-Bailly Consulting (1999).

24. These data are contained in appendices to CEC (2003a).

25. The evaluation report is Goett (1999).

26. The evaluation report is RER (1997).

27. Outside the DSM literature, this is often known as the "Heckman-Mills" ratio.

28. See Dehejia and Wahba (2002); I am grateful to Billy Pizer for bringing this to my attention, and for his advice on the issues discussed in this section.

29. Appendix J of the *Protocols* is "Quality Assurance Guidelines for Statistical, Engineering, and Self-Report Methods for Estimating DSM Impacts."

30. This section again draws upon Vine et al. (2004).

31. The CPUC, in consultation with the CEC, has instituted very aggressive new DSM energy-savings targets (and authorized substantially increased corresponding funding levels) for the publicly regulated utilities, beginning with the 2006–2008 funding cycle; these targets are aimed at achieving a long-run reduction in aggregate per capita electricity consumption in California of 0.3-0.4% per year. The ambitiousness of this goal was noted by Messenger (CEC 2003b): "A sustained reduction in per capita electricity use has never been achieved by an advanced economy in modern times."

32. In June 2005, Governor Schwarzenegger announced that California's official policy would be to reduce emissions to 2000 levels by 2010, to 1990 levels by 2020, and to 80% below 1990 levels by 2050. A first comprehensive assessment of potential policies, programs, and measures for reaching these goals was released in April 2006 (CalEPA 2006). The newly expanded energy-efficiency policies and programs noted above are explicitly regarded by the state as a central element in the developing overall strategy to meet these emissions reductions goals.

33. The topic areas to which the framework is addressed are: impact evaluation, and measurement and verification approaches; process evaluations (i.e., of the operation of programs themselves); information and education program evaluation; market transformation program evaluation; non-energy effects evaluation; uncertainty; sampling, and cost-effectiveness analysis.

34. TecMarkets Works Framework Team (2004), p. 133–134.

9

Concluding Observations
What Can We Learn from the Case Studies?

Richard D. Morgenstern and William A. Pizer

This book began with the question of whether voluntary environmental programs work—that is, whether they actually improve environmental outcomes. Breaking the question down further, we have asked: Quantitatively, how large are the gains? Are there significant differences between energy- and nonenergy-related programs? How do participation incentives, as well as the process for setting goals, affect outcomes? How convincing are alternate approaches to baselines and evaluation? We now want to re-examine these questions looking across the seven case studies in this volume.

A natural starting point for our examination is to summarize what the authors conclude about the performance of the individual, voluntary programs. Do the authors find that participating firms (or households in the case of the California Demand-Side Management program) achieved better environmental results than the determined baseline? Is the baseline credible? If so, and particularly if the results are shown to persist over time, was participation in the voluntary program the principal motivating factor or might other planned or coincidental factors play important roles?

Our value as editors comes from added insights that arise from looking collectively at these studies. Specifically, we can compare the institutional and regulatory contexts of different voluntary programs, including the extent to which specific carrots and sticks may have influenced program participation and effectiveness; the importance of using voluntary approaches to address energy as opposed to toxics-related issues; and the key methodological concerns relative to program design and evaluation that cut across the studies. Finally, we make a bold attempt to compare the effectiveness of the seven programs on a quantitative basis. That is, we attempt to draw some general conclusions about the likely range of effects from voluntary programs based on the range of case study results, with particular attention to transient versus long-lived effects.

Conclusions of the Individual Authors

In this subsection, we briefly describe the different programs and review the key findings and observations drawn by the individual case study authors.

The 33/50 Program

The U.S. Environmental Protection Agency's (EPA) first voluntary program, 33/50, was established in 1991 amid rising interest in finding a quick, cost-effective, relatively noncontroversial approach to address concerns about toxic releases. Focusing on 17 high-priority chemicals reported to the Toxic Release Inventory (TRI), the program emphasized pollution prevention as an environmental management technique. The 33/50 name derives from the program's goal of a 33% reduction by 1992 and 50% reduction by 1995 below a 1988 baseline. According to case study author Khanna, the specific goals were first proposed by EPA Administrator Reilly in 1990.

At the outset, industry leaders expressed concern about possible mandatory requirements and praised 33/50 as a preferred alternative to regulation. EPA chose the 1988 baseline in part to permit companies to take credit for activities already underway and thus not to penalize them for cutting emissions prior to 1991. In fact, more than a quarter of the recorded reductions took place before the official start of the program.

In early 1991, EPA invited the top 600 companies, which accounted for two-thirds of total 33/50 releases in 1988, to participate in the program. Meetings were held with top corporate executives, trade associations, and others to motivate action. Subsequently, second and third rounds of invitations were sent to smaller firms and outreach also was conducted. Sixty-four percent of the top 600 firms agreed to participate, far more than the 10–18% of the invitees from later rounds. The relatively flexible goals could be met by reductions to air, water, or land releases of any of the listed chemicals. Timetables for individual firms also were flexible and there were no requirements for the use of specific abatement methods.

Although some of the reductions clearly were driven by mandatory provisions of the Montreal Protocol and the 1990 Clean Air Act Amendments, covered releases declined considerably between 1988–1995, well in excess of the established goals. The author cites several sophisticated studies, including her own analysis of the chemical industry, which find that the program had a statistically significant, negative impact on releases of the 17 chemicals targeted by 33/50. A major innovation of these studies was to model explicitly the relationship between a firm's decision to participate and its environmental performance once in the program. These statistical analyses were complemented by case studies documenting that once signed up for 33/50, some participants made significant efforts, often at great cost, to reduce their releases.

In contrast to these affirmative results, Khanna reports on a recent study that found the program's overall impact on toxic releases to be ambiguous and, in some cases, adverse; joining 33/50 increased emissions relative to nonparticipating facilities in the chemical industry and a number of other key industries. This study specifically excluded the two ozone-depleting chemicals that were being phased-out by the Montreal Protocol. While Khanna draws no firm conclusions on these new results, at a minimum they raise questions about the performance of this program.

Analyses of firms' motivation for cutting their toxic releases suggest that the desire to differentiate themselves from rivals, to garner positive publicity, and to respond to perceived regulatory threats were important in stimulating program participation. Some companies simply welcomed formal recognition for efforts already underway. Not surprisingly, a number of studies found that firms with greater consumer visibility (e.g., those producing final goods, or in direct contact with consumers), or in industries with higher advertising expenditures per unit of sale, or those operating in states with high environmental group membership were more likely to participate in the program. Other studies found that participation was motivated by the desire to offset adverse publicity or, perhaps, to reduce the frequency of EPA inspections. Interestingly, there were no significant differences in the extent of early reductions carried out by firms signing up in the first round of invitations to join 33/50 versus those invited to join in later rounds. Research also indicates that an environmentally aware public and a credible threat of mandatory regulation if voluntary approaches failed also were important elements, as was the presence of clearly defined numerical goals and mandatory reporting requirements via the TRI.

Japan's Keidanren Voluntary Action Plan on the Environment

Japan's Keidanren Voluntary Action Plan on the Environment was initiated by industry in 1997, just prior to the negotiation of the Kyoto Protocol. It encompasses large enterprises drawn from 58 business associations, including the industrial, electricity, construction, commercial, and transport sectors. Neither small/medium enterprises nor households are part of the Keidanren. Based on 1990 data, Keidanren members represented more than four-fifths of the total greenhouse gas emissions associated with industrial activity and electricity generation and almost one-half of Japan's total emissions.

Under the plan, nonbinding targets were established at the sector level through the individual industry associations that did not apply to individual enterprises. As case study authors Wakabayashi and Sugiyama explain, the plan initially was embraced by industry as a means of demonstrating cooperation with the government on greenhouse gas emissions while, at the same time, it sought to avoid mandatory requirements. At present, they argue, the nonbinding targets are widely recognized as commitments with which industries are to comply.

Although the government has provided tax and other incentives to help address various environmental problems in Japan, no specific economic incentives are available to Keidanren members to help achieve the Kyoto targets. At the same time, the existence of relatively stable, long-term institutional relationships among the government, individual industries, and the Keidanren are critical to the success of the plan. Because the Keidanren has considerable influence on the operation of Japan's industrial policy, it can help to ensure the performance of its members since noncompliance would undermine the multifaceted benefits of the long-term relationships.

Overall, Wakabayashi and Sugiyama note three factors that seem to be motivating industry to comply with the plan: 1) the cooperative relationship between the Keidanren and companies; 2) threats of mandatory policies such as a tax or cap-and-trade schemes; and 3) awareness of private companies' social responsibility. As the authors suggest, some of the motivating factors may be unique to the special relationship that exists in Japan between the government and business and may not be readily applicable to other nations.

In terms of absolute emissions, Keidanren members are committed to stabilizing their collective greenhouse gas emissions at 1990 levels by 2010—a goal for which they are now on track. The key question, confounded by a slowdown in GDP growth during the early years of the plan, is whether this goal is significantly different from business-as-usual. Evidence presented in the case study shows that the emissions of the participating industries have fallen by an average of 0.62% per year over the period 1997–2004. This contrasts with annual emissions increases averaging 0.55% for the seven preceding years. A 2005 follow-up reported by industry found that for 35 of the individual associations that participate in the plan—mostly in the industrial and electricity sectors—emissions were slightly below base-year levels. Were it not for certain unplanned nuclear shutdowns, the industry estimates that greenhouse gas emissions would have been 2.4% below base-year levels.

There are a number of reasons why it is difficult to evaluate whether or to what extent the goals of the plan exceed business-as-usual. Because of its wide coverage, it is not possible to identify a comparable set of nonmember industries (or firms) against which to compare the environmental performance of the Keidanren members. Time-series analysis is hindered by the absence of historical data on emissions by Keidanren members. However, some evidence of the seriousness of the commitment is seen in the actions taken by specific industries. For example, power companies have announced their intention to acquire Kyoto credits from other nations in case they do not meet the targets. The steel industry has expanded efforts to increase efficiency and reduce raw-material waste, in some cases with governmental support. In 2002, the Keidanren established its own evaluation committee to oversee the performance of its members. To date, it has issued a number of recommendations to, among other things, avoid double counting and adopt more standardized assumptions. At the same time, the nongovernmental organization community has raised various concerns about the stringency of the targets and the lack of consistency and transparency in the process.

In sum, the authors present somewhat of a mixed picture on the environmental success of the plan. Even though it is labeled as voluntary, there are clearly some subtle, semi-binding aspects to it. While at this point emissions of the covered industries are on track with the stated goals, there is uncertainty about the future. Moreover, despite the difficulty in documenting the situation, there is some concern that the stated goals may not differ greatly from business-as-usual.

UK Climate Change Agreements

The United Kingdom was an early and strong supporter of the Kyoto Protocol and has adopted a proactive position on climate change, both domestically and internationally. In various white papers, the government set out a far-reaching strategy for achieving and then moving beyond the Kyoto targets and timetables, aiming to reduce CO_2 emissions by 20% below 1990 levels as an interim step in reaching a long-term goal of a 60% reduction by about 2050. In laying out the context for the case study, authors Glachant and de Muizon note the large-scale substitution of gas for coal in both the electricity and industrial sectors that occurred in the early and mid-1990s, driven largely by air pollution issues and the privatization and restructuring of the electricity sector.

In 2001, the UK Government established voluntary, quantified, climate change agreements (CCAs) with 48 sectoral associations in the industrial, commercial, and public sectors as part of a complex policy mix involving an energy tax, a climate change levy, and an emissions trading system. One of the stated goals of the £5–10 ($9–18) tax per ton of CO_2 emissions was to transfer the burden of taxation from employees to CO_2 emissions (the double dividend argument) and, simultaneously, to shift the burden away from energy-intensive industries. The mechanism used to accomplish these goals was to reduce various employment-based taxes alongside implementation of the CO_2 levy. This, in turn, was combined with the use of the CCAs, which then exempted participating firms from 80% of the levy.

When first taking on a CCA, firms could choose either intensity-based or fixed targets (most chose the former) expressed in terms of either energy use or carbon emissions. Overall, the CCAs cover about 12,000 individual sites—virtually all those eligible—representing almost 44% of total UK industry emissions.

Compliance with the CCA could occur via reductions in energy use or by the purchase of emission rights in the recently established pilot emissions trading program. Because most companies were reluctant to enter into a scheme based on collective compliance, several options were offered. The most popular one involved two phases wherein verification is first made for the sector as a whole. If the sector's target is met, all firms are deemed to be in compliance. If the sector target is not met, the performance of individual firms is then examined.

Citing a government-sponsored study, the authors argue that the selected targets were considerably below business-as-usual. However, as they also note, the same study indicates that the targets actually were less stringent than the expected levels with the full CO_2 tax in place; that is, without the CCA tax exemption. Further, the authors reference a study by the Association for Energy Conservation that challenges the initial finding that the targets were significantly below business-as-usual levels. Yet, there is no clear basis on which to judge the merits of the competing assessments.

Regardless of how the targets were set, aggregate emissions during the first two years of implementation were well below them. The case study authors then develop an evaluation methodology to assess the contribution of the CCAs to the implied emissions reductions based on the observed prices for emissions credits and on the behavior of individual firms. Although a government-sponsored study finds widespread compliance with the CCAs, given the low observed credit prices and the relatively small number of transactions, the authors conclude that the CCAs only were modestly effective in encouraging reductions beyond business-as-usual. At the same time, they observe that the use of permit prices as a measure of the effort required to meet the targets may understate the true effectiveness of the CCA if, in fact, the voluntary agreements were effective in reducing organizational and informational barriers that may have hindered the operation of profitable abatement actions.[1] Unfortunately, no specific evidence is available to test this hypothesis.

Denmark's Voluntary Agreements on Industrial Energy Efficiency

Beginning in 1996, the Danish Energy Agency established voluntary agreements on energy efficiency as part of a set of revenue-neutral CO_2 and other green taxes imposed on the industrial, trade, and service sectors. After an initial phase-in period, the full tax rate was 13.3 Euros ($18) per ton of CO_2. Lower rates were applied to energy-intensive firms and those most vulnerable to foreign competition. Virtually 100% rebates were given to energy-intensive firms if they entered into a voluntary agreement on energy efficiency with the energy agency. The voluntary agreements thus were considered complements to the tax scheme. If companies failed to follow through on their agreement, there was an explicit sanction: they had to repay the rebate in full. As case study authors Krarup and Millock note, the rebate effectively lowered taxes on industry to the point where they were only about one-third as much as those on households.

Although the voluntary agreements did not involve quantitative targets, rebates were initially conditioned on the completion of verified energy audits and the implementation within three years of all measures estimated to have a payback that exceeded given criteria. If no energy-savings projects were identified in the audit, the company was considered to be energy efficient and need not carry out new projects to qualify for the rebates. Participating firms also were eligible for a subsidy of up to 50% of the cost of the audits. The investment criteria for the efficiency measures (generally a four-to-six-year payback) were somewhat more relaxed than those typically applied to private investments. Measures to be undertaken included energy-savings projects, so-called special investigations, and energy-management systems such as those specified in ISO 14001. Environmental and consumer groups only had limited involvement in the voluntary agreements.

Issues arose from the outset concerning the high administrative costs of the program. As a result, the voluntary agreements were revised in 2000 and again in 2003, including expanding coverage to additional industries, dropping the requirement for the verified energy audit, and emphasizing adoption of energy-management systems as opposed to audit-based requirements.

Government-sponsored evaluations of the impact of the agreements suggest a 2.6% reduction in energy use over the period 1996–1999 and 1.9% reduction over the period 2000–2003. Other studies estimate somewhat higher reductions, on the order of 3–8%, based on an econometric analysis of a panel of both participants and nonparticipants. Given the small number of participants in the sample (~2%), however, the authors interpret these quantitative results with considerable caution. Qualitatively, company managers in a number of interviews generally thought the program had favorable effects on energy use.

Overall, the authors argue that the true savings of the program likely are quite modest. Importantly, the evaluations also found that the most profitable investments were realized in the agreements implemented in the early years of the program (prior to 2000). In the future, as the EU Emissions Trading Scheme is implemented, the authors see only a very limited role for this type of voluntary agreement in Danish greenhouse-gas-mitigation policies.

The German Cement Industry

In 1995, the Federation of German Industries, a group of 16 industrial associations representing major sectors of German industry, voluntarily issued the "Declaration of German Industry on Global Warming Prevention" (GGWP), which called for voluntary reductions in specific fuel consumption (e.g., energy per unit of output) of up to 20% below 1987 levels by the year 2005. The initial commitments undertaken by the German industries did not involve any government-provided incentives nor were they accompanied by threats of future regulation. The initial GGWP simply was a unilateral commitment by the nation's principal industries to reduce their emissions. By the year 2000, five years in advance of the target date, most of the commitments of the individual industrial associations already were fulfilled, an indication, in the view of case study authors Böhringer and Frondel, that the targets were not very ambitious. Subsequently, as the result of pressure by the government and the desire of industry to avoid mandatory requirements, the GGWP goals were made more stringent: a 28% reduction in energy-related specific CO_2 emissions (e.g., emissions per unit of output) below the new base year 1990 by 2012. Moreover, the declaration was joined by other industry associations.

For purposes of evaluating the effectiveness of the GGWP declaration, the case study authors emphasize the importance of establishing a credible baseline. For that reason, they single out the cement industry for detailed analysis because it is the only one among the 19 industries now in the German Federation for which sufficient historical data are available to compare the CO_2 emissions of the industry following development of the GGWP to emissions in prior years.

Using data for the period 1974–1995, which begins after the 1973 oil shocks and ends prior to the start of the GGWP, the authors estimate annual fuel efficiency improvements of 0.63% in the cement industry. By extrapolating this trend forward and comparing the observed to the predicted levels of fuel efficiency, Böhringer and Frondel conclude that the energy- and emissions-reducing activities "have not gone much beyond good intention." Indeed, they find that the carefully monitored performance of the cement industry during the period of the GGWP declaration does not deviate significantly from business-as-usual, noting that the margin of error on the forecast business-as-usual baseline is ± 5%. More generally, the authors argue that because of asymmetric information between the industry and government, it is quite difficult for outsiders to decide whether the commitments made by the cement or other industries push beyond the levels that would have occurred without the voluntary agreements in place. That is, the impossibility of observing the counterfactual, along with the absence of pre-declaration data that could be used to develop a credible assessment, make it relatively easy for industry to declare the effort a success. Performance monitoring, by itself, they argue, is insufficient to gauge the effectiveness of the program.

As regards policy recommendations, the authors call for the use of firm-specific targets as opposed to aggregate, industry-wide agreements. Further, they argue that such agreements negotiated between firms and regulators, rather than being unilaterally set by industry, have a far better prospect of encouraging emissions reductions beyond business-as-usual.

Climate Wise

Climate Wise is a voluntary program with the nonutility industrial sector developed by U.S. EPA to encourage the reduction of CO_2 and other greenhouse gases. Originally established in 1993, Climate Wise remained in operation until

2000, when it was renamed and placed under the agency's Energy Star umbrella. Unlike EPA's well-known technology-based programs (e.g., Green Lights) that require the adoption of specific technologies, Climate Wise members have the flexibility to adopt whatever technologies or strategies they choose to reduce their emissions. The requirements simply are that a participating firm develop baseline emissions estimates of its greenhouse gases for any year since 1990, self-designate forward-looking emissions reduction actions, and make periodic progress reports. To ensure that the proposed reductions were substantial in nature, EPA provided a checklist of major actions, such as specific boiler modifications, waste heat recovery systems, and others. Firms were strongly encouraged to select at least some of their proposed actions from this list. Participants also were encouraged to report their progress to the U.S. Department of Energy through the 1605(b) registry program and to provide a copy of the completed form directly to EPA.

EPA offered several kinds of technical assistance to participating firms, including a guide to industrial energy efficiency, various government publications on energy efficiency, and free phone consultation with government and private-sector energy experts retained as consultants by the agency. The EPA also set up an annual event open to the public to recognize the performance of outstanding Climate Wise participants. As part of these events, a series of workshops were held that allowed participating firms to exchange experiences about their efforts to improve industrial efficiency and reduce greenhouse gas emissions. Informal reactions from agency staff and industry representatives suggested that these workshops were seen as quite valuable by the participating firms. At its peak, Climate Wise had enrolled more than 600 industrial firms covering several thousand facilities nationwide.

While EPA has published estimates of emissions reductions associated with Climate Wise ranging from 3 to 20 million metric tons of greenhouse gases, these estimates have been criticized by other researchers. In particular, they expressed concern that the extremely wide range of activities covered by Climate Wise makes the program's role in the decision to undertake these activities somewhat questionable. The principal goal articulated by case study authors Morgenstern, Pizer, and Shih is to try to tease out the unique contribution made by Climate Wise apart from other factors that may have contributed to the reported emissions reduction by the program participants.

To conduct their evaluation, the case study authors obtained access to confidential, plant-level data files for the manufacturing sector collected by the U.S. Census Bureau. These files contain technical and economic information (e.g., data concerning location, output, energy expenditures, and other relevant variables) on individual plants in multiple industries. The authors first use this data to create a matched set of nonparticipants, where each Climate Wise participant is paired with another facility with similar observable characteristics. The authors then compare the performance of Climate Wise participants and non-participants.

Although a number of important caveats apply, the principal result reported by the case study authors is that Climate Wise appears to have had little to no effect on fuel use, while slightly increasing demand for electricity. Comparing the change in fuel and electricity expenditures before and after a facility joined the program and then across participants and nonparticipants over the same horizon—a difference-in-differences approach—their best estimate is a temporary (1–2 year), 3% decline in fuel use and 6% increase in electricity use associated with joining the program. Given a roughly ± 5% confidence interval, the effect on fuel use is not consistently significant.

Given the counter-intuitive result for electricity, the authors discuss a number of reasons why electricity use may have risen. On an econometric level, the variable used to match participants and nonparticipants on growth—changes in the value of shipments—may be inadequate, and one might see faster growth—and therefore increased electricity use—among participants. On a substantive level, it may be that firms are making choices to increase electricity use to reduce direct CO_2 emissions. Finally, the fuel expenditure variable may not be a good proxy for greenhouse gas emissions. Although the focus of the Climate Wise program was on energy efficiency and the reduction of CO_2 emissions, a number of firms in the chemical industry proposed reductions of nitrous oxide and some breweries proposed reductions of methane. Other firms may have pursued fuel-switching projects that would not show up as a decrease in expenditures.

The one result that is clear, regardless of the effect on electricity, is that all of the significant effects vanish after two years, suggesting any program consequence is temporary.

Residential Demand-Side Management Programs in California

Beginning in the 1970s, at the instigation of the regulatory authorities, California electric and gas utilities sponsored programs to promote the adoption of energy-efficient technologies and energy-conserving behavioral practices. These programs included general and site-specific information programs and financial assistance. Case study author Sanstad focuses on information-only programs for single-family dwellings conducted by investor-owned utilities in the 1990s, wherein the utility would assess customers' energy consumption patterns, equipment holdings, and other energy-relevant characteristics and, at the same time, offer advice on conservation practices and potential efficiency investments. Programs where explicit financial assistance was offered are not considered. Unlike the other programs considered in this volume, the focus in this case study is on the residential sector rather than the industrial sector.

The literature on programs of this type has emphasized the need for undertaking ex post performance measurements in addition to ex ante calculations and for addressing the issues of counterfactual baseline and self-selection. The author notes that California has in recent decades established regulatory requirements for demand-side management (DSM) program evaluation to address these issues. He reviews the results of evaluations reflecting these requirements conducted in the mid-1990s on programs undertaken by three of the state's four large, publicly regulated, investor-owned utilities: Pacific Gas & Electric, Southern California Gas, and Southern California Edison.

For each of the three programs, evaluations compared energy use before and after the program was implemented and applied one or more methods to address selection bias. Nonetheless, the author finds that discerning the true impact of the programs is technically and practically challenging even with state-of-the-art methodological controls. He also notes substantial differences in technical sophistication across the three evaluations. After carefully reviewing the results, he concludes that at least two of the three programs yielded energy savings on the order of several percent that would not have occurred in their absence. This magnitude is consistent with previous findings of the savings accruing from DSM programs such as these that do not include financial incentives.

In the more detailed Southern California Edison study, savings are reported to be driven principally by changes in the behavior of households (e.g., improving maintenance of appliances or discontinuing use of secondary refrigerators) rather than by the installation of new equipment. At the same time, as the author notes, "while vaguely aware of the energy benefits of the recommended actions, [customers] . . . do not always act on this knowledge until it is suggested by an expert." This implies that a key barrier to action by homeowners may not be information per se but information from an authoritative source.

Observations Across Studies

In Table 9-1, we attempt to quantitatively summarize and compare the effect of these different programs based on the work by the authors of the individual chapters. At first glance, the estimated effects range from zero in the case of the German cement industry to 28% in the case of the 33/50 program. Most of the estimates, however, are in the 5–10% range. How can we understand these results?

Media and Activity

As noted early on, a key difference between the 33/50 program and the others examined in this volume is its focus on toxics rather than on energy or energy-related greenhouse gas emissions. Toxic emissions differ from energy and

TABLE 9-1. Quantitative Comparison of the Effect of Voluntary Programs on Behavior

	Quantity measured	Estimated effect	Scope	Baseline	Comment
33/50 Program	Aggregate toxic releases	28%	Participating chemical facilities	Nonparticipants with self-selection model	Some evidence that effect is reversed when ozone-depleting substances excluded.
UK Climate Agreements	GHG emissions	9%	Participating industries	Negotiated forecast	Baseline criticized; considerable over-achievement.
Danish Energy Efficiency Agreements	Energy use	4–8%	Participating facilities	Nonparticipants	Estimate based on 60 participants.
German Cement Industry GWP Declaration	Energy per unit of cement	0	German cement industry	Econometric forecast using historic performance	Baseline error band is ± 5%. 2005 target achieved by 2000.
Japanese Keidanren	CO_2 emissions	5%	Participating industries	Keidanren forecast of 2010 business-as-usual	Basis of business-as-usual estimate unclear.
Climate Wise	Fossil energy expenditures	3%	Participating facilities	Matched nonparticipants	Electricity expenditures estimated to rise 6%. Margin of error is ± 5% and both effects vanish after 1–2 years.
California Demand-Side Management	Natural gas and electricity demand	2–4%	Participating households	Nonparticipants	Covers three programs; some evaluations more carefully matched nonparticipants/controlled for self-selection

energy-related greenhouse gas emissions in three important ways. First, toxics typically are a local or regional pollutant, while greenhouse gases are global. Second, toxics can have a direct, acute effect on human health. They also can have long-term, chronic effects, such as cancer and heart disease. Meanwhile, greenhouse gas emissions accumulate for many years, affect the climate, and thereby impact ecosystems and overall human well-being over the longer term. Finally, with no practical opportunity for end-of-pipe abatement, reductions in energy-related greenhouse gas emissions often amount to reductions in energy use itself. Given the underlying positive price on energy, there is always an incentive to reduce energy use. Toxic emissions, meanwhile, are often an unpriced industrial byproduct whose existence was widely ignored until the 1980s and 1990s.

What does this mean for program effectiveness? It should not be surprising that the potential effect of a voluntary program addressing toxic releases is higher—in terms of percentage reductions—than that of a voluntary, energy-related program. Each of the three reasons cited above implies a greater incentive and potentially lower costs for reducing toxics. As noted by Khanna in Chapter 2, public recognition and goodwill often is a key motivation for corporate action in a voluntary program. Toxics arguably offer a greater reward in this area because of their more localized consequences. Reductions in mercury releases, for example, provide far greater public relations benefits for firms than lower natural gas usage. This arises in part because households are more familiar with energy issues, making energy use and savings less dramatic, and in part because climate change affects society collectively and globally as opposed to the more localized effects of most toxics.

In addition to larger benefits, firms also may find lower costs associated with efforts to reduce toxics versus additional energy efficiency. Why? Firms always have an incentive to seek out reductions in energy use because such reductions save money, especially when energy prices are high (as in recent years). The added incentive of a voluntary program to reduce energy use on top of existing energy prices seems less likely to turn up many new opportunities. In contrast, there is no obvious reason firms would have previously sought reductions in toxic emissions until they were alerted to the problem. In this way, reduction in toxic releases might represent a new arena of inquiry, with the consequent possibility that such inquiry might yield significant, low-cost reduction opportunities.[2]

More generally, it is important to consider how the kind of environmental hazard and mitigation activity addressed by a voluntary program will affect both the incentives for participation and action as well as the likelihood that achievable, affordable opportunities will be uncovered. Programs that address local, popular concerns likely will resonate more with companies and their constituents. Similarly, programs that focus attention on a relatively new issue may uncover more opportunities than programs that draw attention to areas of existing concern.

Incentives for Participation and Action

The preceding discussion points out that the nature of the environmental problem addressed by a voluntary program creates different incentives for both participation and action. More generally, voluntary programs are structured with different incentives, ranging from information benefits and subsidies, to the threat of mandatory regulation, to the coercive effect of strong trade associations, to specific relief from otherwise burdensome taxes, as in the case of the UK and Danish programs. It is, arguably, the nature and magnitude of the different incentives that creates the strongest distinction among voluntary programs.

Certainly we see differences in terms of participation. The Climate Wise program, with relatively weak incentives, had relatively low participation rates. Meanwhile, the UK agreements and the Keidanren program in Japan, with strong incentives, had almost universal participation. More interesting is the unilateral, incentiveless German program, which covered entire industries, and the Danish program, which offered significant tax incentives, yet saw only partial participation.

To understand these differences, we also have to consider the requirements of/for participation. In the Danish program, there were real costs to participating firms associated with the required audits (estimated at 17,000–33,000 Euros per facility). In the German program, there were no costs to participants. Taking such costs into consideration, the pattern of participation makes more sense.

The effect of incentives is perhaps even more interesting when we turn to the actual results. All of the energy-related programs have effects in the 0–10% range and all but the UK program suggest an effect closer to 5% or less. This is true regardless of how large or small the incentives for participation or action might be. This 5% estimate was also highlighted in Chapter 8 as being consistently observed across a variety of residential energy programs.

Methodology and Baseline

As we noted in Chapter 1, measuring the effect of these programs requires a baseline. The ideal baseline would come from a randomized experiment, with participation in the voluntary program assigned by pure chance to one group of firms or households and the remaining nonparticipants serving as controls. Absent such an experiment, analysts have used one of two general approaches: either a baseline forecast based on historical data and/or other information or a group of nonparticipants serving as a control group.

Among the programs surveyed in this volume, three use a forecast baseline to estimate program effects and four use a control group. Within each of the two general approaches, we see a variety of implementations. Both the UK and

Japanese programs use baselines collaboratively developed by business and government. The UK program, however, was criticized in part because so many industries overachieved their targets and did so with minimal price incentives. While the Japanese program was subject to less criticism, the basis of the forecast remains unclear.

Böhringer and Frondel take an entirely different approach to providing a forecast baseline. While the commitment itself did not provide a baseline estimate to evaluate the target (which was achieved in half the time originally planned), the authors simply extrapolated past trends. They demonstrate that the commitment itself represented nothing more than a continuation of historical trends (though this trend itself had a ±5% margin of error). They conclude that the target did not represent an effort beyond business-as-usual, something they attribute to the fact that the industry came up with the target unilaterally instead of negotiating it with the government or other stakeholders.

It is interesting that a similar argument was leveled in 2002 when the Bush administration announced its economy-wide target of an 18% improvement in greenhouse gas emissions per dollar of GDP over 10 years. While forecasts at the time suggested a baseline of 14%, critics noted that intensity had improved by 18% over the preceding 10 years.[3] Based on the Böhringer and Frondel approach of simply extrapolating past trends, such a target would be viewed as business-as-usual, even though it represented—at the time—a 4% improvement over the more elaborate forecasts prepared by the Energy Information Administration. Now, as the United States finds itself currently on track to meet the 18% commitment, these issues have arisen all over again, as well as new ones. Namely, even if 14% was a reasonable business-as-usual forecast and 18% a reasonable target in 2002, we would expect larger improvements in greenhouse-gas-intensity with higher energy prices in 2006.

The underlying problem is that BAU forecasts can be outdated quickly by events completely unrelated to a particular environmental program.[4] Where participation is widespread (e.g., industry-wide programs such as the Keidanren and German industry targets), there may be no alternative to such an approach. However, when only a portion of the population participates in a voluntary program, it is natural to look at the behavior of nonparticipants to provide a baseline.

Unfortunately, participation in voluntary programs is not random and, as noted, a typical participant often looks very different than a nonparticipant. In the Morgenstern, Pizer, and Shih study of the Climate Wise program, the authors find that participants are larger and faster-growing than their nonparticipating counterparts. In one of the studies surveyed by Sanstad, participating households had higher incomes and dramatically higher frequency of air conditioning (central and room) than nonparticipants.

Program effects still can be estimated when participant and nonparticipant demographics differ, so long as those demographics are included in the model and are specified correctly. This is the approach taken in the evaluation of the Danish climate change agreements and the 33/50 program, as well as some of the DSM programs just mentioned. An alternative is to more carefully match participants and nonparticipants, as was done in other DSM programs surveyed by Sanstad and in the Climate Wise evaluation.[5] Careful matching can eliminate the demographic differences between program participants and controls. Also, it puts less pressure on the model specification (although, as the Climate Wise study shows, the matching effort can be as important as the outcome specification).

A final concern is that even if a nonparticipant control group looks similar to the participants we want to evaluate, there may be unobserved differences that explain why participants joined and nonparticipants did not. For example, participants simply may be more environmentally concerned, leading them to join and to have better environmental performance but not reflecting a real effect of the program. In this case, we need an explicit (and correct) model to explain how unobserved features—errors—might be related across the participation decision and the outcome. The studies reviewed by Khanna and by Sanstad consider this issue.[6]

Synthesis

Tying these observations together, an interesting result emerges. The context of the program, particularly the additional use of carrots and sticks to encourage and strengthen the effectiveness of voluntary programs, appears to have only a limited effect on the measured quantitative results among participants. While we have only one example, much larger differences seem to be associated with the kind of problem addressed by the voluntary program; that is, toxics versus energy-related activities. Meanwhile, incentives do have a major influence on the degree of participation. Given that the overall impact of a voluntary program is the product of effectiveness per participant and the number of participants, incentives clearly affect the overall impact of a voluntary program.

This observation about effectiveness emerges despite the wide variety of approaches to constructing baselines, an issue that has pervaded this volume. Concern about the UK baseline, for example, makes us wary of the estimated 9% effect, but it is still in the ballpark of the other estimates. Similarly, the estimated zero effect in the German program contains a ± 5% error window. Summarizing, it would be hard to reject the hypothesis that all of the energy-related programs had a 5% impact among participants.

Should this be surprising? On the one hand, we began this exercise with the idea that context would be important; a key element of our case-study design was that author teams were supposed to describe the incentives surrounding the voluntary programs. Some involved tax exemptions and others were not only voluntary but unilateral on the part of business. On the other hand, energy already is a real expense for firms and households. If we imagine an energy demand elasticity of 0.5, a voluntary effort that increased the cost associated with energy use by 10% only would have a 5% effect on demand.[7] Is it sensible, therefore, to imagine a larger effect stemming from a voluntary program?

Even if we see this as a reasonable result, a number of caveats are in order, not the least of which is the limited scope of our analysis. We examined seven voluntary programs among thousands in operation in the United States, Europe, and Japan. Further, there are a variety of softer effects that may be equally if not more important, such as changes in attitude and practices that may pay off in the long run. There are nontrivial margins of error in these evaluations; in some cases, they are not even specified. In other cases, larger questions lurk in the background (for example, inclusion of regulated toxic releases in the 33/50 evaluation). It also is worth noting that a 5% improvement in energy use is not a trivial accomplishment, especially for a voluntary program and especially if participation is fairly broad. Nonetheless, it does suggest certain limitations on such endeavors.

Conclusion

We noted at the outset of this chapter that our discussion of voluntary programs is by no means comprehensive. Yet, we have examined programs in three different regions of the world, with a variety of designs and incentives, and with a range of efforts at evaluation. All but one of the analyses suggested that the voluntary program affected behavior; however, focusing on energy-related activities, the effect was less than 10% and more typically closer to 5%. One evaluation suggested a zero effect (the German cement industry study) but offered an error bound of ± 5%. We note that even in the case of the 33/50 program covering toxics, where a 28% improvement was observed, other studies that excluded substances regulated under the Montreal Protocol and/or domestic regulations found negative effects. A tentative conclusion, therefore, is that voluntary programs have a real but limited quantitative effect, particularly among energy-related activities.

Of course, a 5% reduction in energy use or carbon dioxide emissions is not trivial. Some nations' initial efforts under the Kyoto Protocol amount to roughly that order of magnitude. It also represents potentially billions of dollars in savings. Nonetheless, it represents what appears to be an outer limit on what these kinds of programs can achieve.

At the same time, many of the authors have noted important but hard to quantify soft effects, such as changes in attitudes or management practices that are viewed by participants and stakeholders as significant steps in improving long-term stewardship. While such effects may not show up as immediately quantifiable reductions in targeted emissions or energy use, they may lead to longer-term, broader improvements in environmental performance.

A key theme in the chapters and our discussion has been the methodology used to evaluate programs. An important lesson, then, is the importance in establishing a reasonable baseline. This involves a careful negotiation of a reasonable business-as-usual forecast or, ideally, identification of a suitable control group for evaluation.

There is some evidence that incentives can affect the magnitude of emissions reduction or efficiency improvement. The energy-related programs with the weakest incentives—Climate Wise and the German GWP declaration—had the weakest effects. Those with the strongest incentives—the UK and Danish agreements and the Japanese Keidanren program—had the strongest effects. However, the difference is somewhat suspect, especially because the UK and Japanese programs were evaluated against a forecast baseline. The difference also is small, with all the programs in the 5% range.

We also noted that in contrast to the limited effect on the magnitude of effects among participants, incentives played a significant role in the level of participation, with some programs with larger incentives and lower barriers to participation having near universal participation. Therefore, despite the lack of a large impact on estimated effects among participants, the fact that the pool of participants is larger means that the overall impact (e.g., effect × number affected) is larger as well.

After estimating the quantifiable effects of a voluntary program, a necessary follow-up question should be whether the effect persists over time and whether, as the program continues, future participants will continue to find the same gains. Evidence from both the Climate Wise and the California DSM programs suggests that some initial gains may not persist. Analysis of the Danish agreements indicates that as the most profitable gains were realized and larger firms joined earlier, later gains were smaller.

At the outset, we imagined that this volume might offer advice to policymakers and stakeholders hoping to better design voluntary programs. Tentatively, that advice might look like this: Create incentives to attract firms into a program and put some effort into negotiating reasonable action and providing a baseline for evaluation, but do not press participants excessively to spur heightened effort. Among energy-related programs, the studies summarized in this book suggest only a slightly stronger effect where taxes, further regulation, or industry coercion were threatened. That is, there seems to be only limited evidence that stronger incentives lead to stronger effects. Rather, the possibility that a larger group of participants can be inspired to act implies that whatever effect there is will occur over a larger population. The Böhringer and Frondel study offers some guidance on a lower bound; however, the threshold for action should not be so low as to be meaningless.

An important initial consideration in designing a program must be the environmental media and activities being addressed. If it is a novel and previously unstudied area, there may be opportunities for more significant improvements at low cost; for example, with toxics. At the same time, if it is an arena that already has been carefully scrutinized, as we believe to be the case for energy efficiency, such opportunities are less likely.

Summarizing, we find that voluntary programs can affect behavior and offer environmental gains but in a limited way. By considering the media and activity, as well as the potential incentives that can be brought to bear, stakeholders can make crude assessments of the potential for a voluntary program. A critical step is having a realistic, agreed-upon baseline. In many cases, such programs make sense; when the arguments for mandatory programs are unclear or lacking legal or political support or where such programs will take considerable time to implement, voluntary efforts can play an important role. However, none of the case study authors found truly convincing evidence of dramatic environmental improvements. Therefore, we find it hard to argue for voluntary programs where there is a clear desire for major changes in behavior.

References

Dahl, Carol. 1993. A survey of energy demand elasticities in support of the development of the NEMS. Washington, DC: U.S. Department of Energy.

EIA (Energy Information Administration). 2005. Impacts of modeled recommendations of the National Commission on Energy Policy. Washington, DC: Energy Information Administration.

White House. 2002. Global climate change policy book, addendum. http://www.whitehouse.gov/news/releases/2002/02/addendum.pdf (accessed July 31, 2006).

Notes

1. We return to this idea of "soft" qualitative results later in this chapter.

2. This same line of reasoning partly explains why analyses of multi-gas policies, such as the emissions trading program proposed by the National Commission on Energy Policy, find that nearly half the reductions come from non-energy-related emissions, even though such emissions are a much smaller fraction of the total inventory (EIA 2005).

3. See White House (2002).

4. The initial National Allocation Plans in the EU ETS have been criticized based on over-compliance, raising similar questions about whether the problem is with the baseline forecast or with events that might have occurred after the forecast was made.

5. Sanstad provides an interesting history of the selection issue in the context of DSM program evaluation.

6. The key part of this approach involves finding an instrument, or excluded variable, that only affects the participation decision and not the outcome directly. This allows the identification of an outcome effect arising solely from participation. In the end, the credibility of this approach rests on the arguments for such a variable.

7. See Dahl (1993) for a review of elasticities estimates, suggesting 0.5 is fairly typical for the intermediate to long term.

Index